CIA's ARSENAL OF

PIPES, IMPROVISED AND WEAPONS PENS

ZIPS, PIPES, AND PENS

ARSENAL OF IMPROVISED WEAPONS

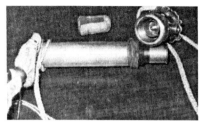
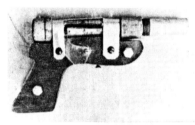

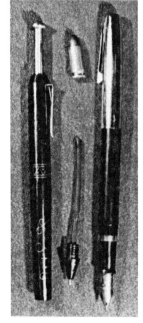

PALADIN PRESS
BOULDER, COLORADO

Also by J. David Truby:
Improvised Modified Firearms: Deadly Homemade Weapons
 (with John Minnery)
Modern Firearm Silencers: Great Designs, Great Designers

Zips, Pipes, and Pens: Arsenal of Improvised Weapons
by J. David Truby

Copyright © 1993 by J. David Truby

ISBN 0-87364-702-5
Printed in the United States of America

Published by Paladin Press, a division of
Paladin Enterprises, Inc.
Gunbarrel Tech Center
7077 Winchester Circle
Boulder, Colorado 80301 USA
+1.303.443.7250

Direct inquiries and/or orders to the above address.

PALADIN, PALADIN PRESS, and the "horse head" design are trademarks belonging to Paladin Enterprises and registered in United States Patent and Trademark Office.

All rights reserved. Except for use in a review, no portion of this book may be reproduced in any form without the express written permission of the publisher.

Neither the author nor the publisher assumes any responsibility for the use or misuse of information contained in this book.

Visit our Web site at www.paladin-press.com

Contents

Introduction	Setting the Target	1
Chapter 1	The Opening Round	9
Chapter 2	Have Another Round	25
Chapter 3	Pipes, Plumbers, and Other Spare Parts	31
Chapter 4	Familial Feud	55
Chapter 5	Asian Genesis	63
Chapter 6	The CIA's Deer Gun	81
Chapter 7	The Fringe's Death Benefits	87
Chapter 8	Barring All Guns	97
Chapter 9	Meanwhile, in the Basement	107
Chapter 10	Under the Umbrella	125
Chapter 11	The Pen Is Mightier	131
	Bibliography	145
	Sources	151

INTRODUCTION

Setting the Target

THE MEN WHO ENGINEERED the Declaration of Independence and the U.S. Constitution, with its Bill of Rights, are among history's geniuses because they created documents with the proper combination of flexibility, timelessness, specificity, and semantic clarity that are as valid today as when they were written. For example, while our founding fathers could not have foreseen television, political commercials, and/or pornographic videos, their wonderful and absolute First Amendment freedoms are as important and strong today as they were more than two centuries ago. Likewise, in 1791, every large muzzle-loading rifle was an assault weapon, yet the amazing clarity and honest truth of the Second Amendment are as legally valid today as we head into our third century.

Yet, here we are, still debating the legal meaning of the Second Amendment. It makes me very sad when I consider those subcaliber Big Bore politicians of today who would disarm our populace by sodomizing that Second Amendment. It makes me sad because, thanks to the First Amendment and because of our nation's love of history, I can read of the absolute genius of true leaders, statesmen, and patriotic heroes like James Madison, Patrick Henry,

Samuel Adams, Thomas Jefferson, and the others who pledged their lives, fortunes, futures, plus something that's missing today from too many of our national politicians, a sense of sacred honor, to found our nation dedicated to individual liberties.

This puts me in mind of one of the more important grievances that sparked our war for independence from British rule. The British tried to confiscate private arms despite the English Bill of Rights of 1689, which gave all citizens of the king, including colonists, the right to keep and bear arms for self-protection.

Obviously, the absentee British government's illegal effort to disarm the colonies was an attempt to strengthen central political control. I don't think the term "police state" had been invented then, but that scenario sure fits today, only our absentee government now sits in the District of Columbia.

The British plan was to disarm the population, then follow with soldiers illegally entering homes and shops to confiscate "contraband" and "seditious literature" and to "punish the treasonable," according to the collected letters of the Colonial British governor, Gen. Thomas Gage.

According to Oliver Wendell Holmes, constitutional scholar and former chief justice of the United States Supreme Court, in a speech to University of Virginia law students:

> Our war with England was a just cause, for it meant our independence from increasingly harsh military, not civilian rule. Our war was fought against an unjust and cruel system that mocked the civilian rights guaranteed to all subjects of the King, including our colonial ancestors . . . We fought a standing army, some of them foreign mercenaries, for our right to be free and independent. We fought the King's standing army with our volunteer civilian army and with our own free militia armed with the firearms the King would have had his soldiers confiscate.

One of the earliest modified weapons in American history, a combination plow/cannon, received a U.S. patent in 1845 and was actually in production for about six months. (Courtesy of Gridley McGeary.)

Thus, this first experience with governmental confiscation of firearms was the spark that moved us closer to an independence that was won not by debate and words, but by honor, courage, faith, hope, and sacrifice by free men and women. According to most historical scholars, it was our ancestors' fear of illegal government control that led the framers of our Constitution to include a very simple and clear written guarantee that no government would ever again intrude upon the rights of each and every free individual citizen to keep and bear private firearms.

When delegates from the 11 original states that ratified the Constitution in 1789 called for amendments for protection of individual rights and liberties in 1791, five states asked for the right to bear arms, five requested a free press, and three requested freedom of speech. With that wonderful genius I mentioned earlier, our founding fathers realized the importance of all individual liberties and freedoms; hence, our Bill of Rights. One of those rights is the absolute right to bear arms.

Despite convoluted pleas from misguided or politically ambitious politicians or the unilateral, bureaucratic, and possibly illegal interpretations of the Bureau of Alcohol, Tobacco and Firearms (ATF), there are no exceptions to the Second Amendment's very clear, simple, declarative language. There is nothing that says "except for assault rifles, firearms having magazines over seven rounds, firearms that look different, or firearms meant only for sporting use." To declare anything except the original directive language of the Second Amendment is absolutely wrong.

What to do about it? There are certain options, all of which I leave open for study and cogitation by the reader. Perhaps we are nearing the point where some of the other language, e.g., revolutionary changes in government, contained in our founding fathers' documents may be appropriate. We are already at the point where there are not enough police to protect us from the evil things that lurk out there right now.

Our obvious first line of defense against these destroy-

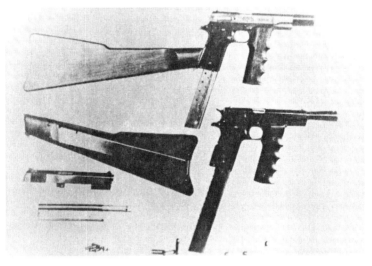

The very rare Swartz conversion of the .45 pistol into a full automatic, stocked weapon. William L. Swartz, a Colt Industries engineer, patented the conversion in 1936, and the army tested prototypes in 1940. This is the only known photo of the only known prototypes. A .22 conversion unit is also shown. (Courtesy of J. David Truby.)

INTRODUCTION: SETTING THE TARGET • 5

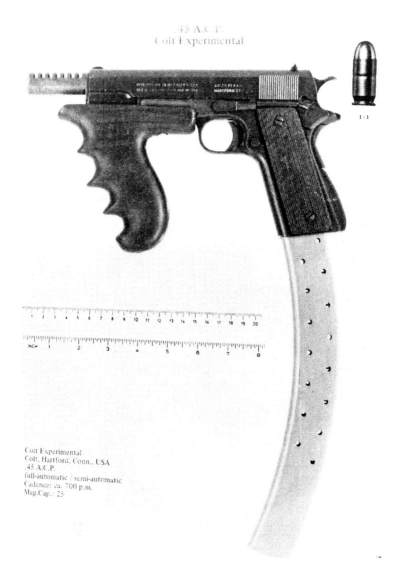

This experimental M1911A1 was converted for full automatic fire at 700 rounds per minute using an oversize, 25-round magazine. (Courtesy of Jack Krcma.)

A German MP44 with several modifications, including the case deflector. (Courtesy of Fred Rexer, Jr.)

One of "Carbine" Williams's first homemade rifles, produced with milling machine or lathe from salvaged parts and an old piece of log. (Courtesy of J. David Truby.)

ers of the Bill of Rights is a good offense. Let's fight them in the legislatures, the Congress, and in the statehouses and the White House. Let's fight for our rights. But let's also show them that from a technological and psychological standpoint, gun control doesn't work for free people anywhere, anytime.

For these reasons, the production of improvised firearms must be part of the curriculum in every free person's study of combat arms training. Personal knowledge and experience in building and operating improvised firearms builds not only a strong confidence in oneself, but also in the potential of these weapons as tools for continued life and freedom in our never-ending war against tyranny.

While small arms technology has come a long way since the first book in this series was published, improvised/modified firearms remain relatively uncomplicated and basic (e.g., a fully automatic weapon is far easier to design and build than a semiautomatic one). Today, all of the improvised/modified designs remain well within the accomplishment of the mechanically unskilled citizen who does not have access to firearms through other means.

Famed mercenary William Morgan in Cuban militia uniform with his modified .30-caliber semiautomatic cut-down rifle fed from a 50-round drum magazine. (Courtesy of Robert K. Brown.)

Suppose there were more restrictions placed on private firearms ownership and the Second Amendment was further sodomized by antigun government forces. Would crime disappear? Would America become a kinder, gentler nation? Would poverty, fear, and hatred vanish, too? I ask these silly rhetorical questions simply because there are probably simpletons of all sorts out there who believe in that nonsense.

But given the fact that such rhetorical mutterings are just that, we are left with another, very real question: If firearms are outlawed, will they just go away? Will people agree to this prohibition, and, if not, what will they do about it?

That is what this book is about.

This book is organized as follows, with four introductory chapters explaining the genre, anatomy, and technology of improvised/modified firearms, followed by five chapters on their distribution and use. The final three chapters discuss the most common types of improvised/modified firearms. As always, myriad photographs and drawings accompany the text, which I have kept on a technical/historical track, avoiding polemic except where necessary to make a point.

For example, I know that gun owners dislike being stereotyped as drooling, redneck, racist, knuckle-walking conservatives. Outdated and unimaginative terms like "liberal" and "conservative" should not even apply to gun control issues. I like the words "free person" and "would-be slave" instead. Beyond that, this issue should not pit pro-gun folks against each other because of race, religion, education, politics, or personal beliefs. An illustrative insight on this issue came as a letter to the editor from a reader of a newspaper in California. In part, it read:

> . . . I am sick and tired of hearing every problem in this country blamed on the Liberal Left . . . I know many diehard Reagan-Bush conservatives who would like to outlaw all private ownership of any form of gun . . .
>
> I am a lifelong liberal, Kennedy Democrat, homosexual, and gun owner. I believe the Second Amendment was written not only to protect the country from invasion, but also to protect the country from an internal takeover . . .
>
> If religious fascists come to get this homosexual, I will defend myself and thank the National Rifle Association for protecting my Second Amendment right to possess a weapon to protect myself.

I am a lifelong antipolitician, populist, combat vet, personal supporter of JFK/RFK, heterosexual curmudgeon, and gun owner. Tell you what, Mr. Gay Gun Owner, if those demented yahoos come for you, count on me. I'll face them with you.

1 CHAPTER ONE

The Opening Round

DESPITE MODERN WARFARE'S COMPUTER-GAME battle scenes shown via U.S. government censors' tightly controlled propaganda telecasts, war is still combat between individuals armed with their own personal weapons, usually a rifle of some sort. Terrorism, another form of deadly diplomacy, also uses hand and shoulder firearms as the basic arsenal.

Despite the truth of the above philosophy, for years nations have sent soldiers and agents into the field ill-prepared to engage in combat with their enemies. Take away a soldier's basic weapon and he or she too often becomes defenseless. To the basic grunt, courses in unarmed combat plus escape and evasion are the best alternatives offered. It's not enough.

Ways must be found to put other weapons into our combatants' hands. Airdrops or submarine landings of arms may not always be available. Plus, there are other problems. For the field agent, personally smuggling arms is almost out of the question now, due to the levels of international security brought on by terrorism. Further, the discovery of an individual in possession of weapons prejudices not only his or her cover, but also personal freedom and/or life. That gun control is so absolute in so many nations has created a vast global police state where governments do not trust their own citizens with private firearms. That people have allowed politicians, bureaucrats, fascists, and dictators of all political stripes to disarm them does not speak well to the collective intelligence of these citizens. That those who are permitted the illusion of free elections continue to support these police-state rulers defies human intelligence and logic even further.

Thus, the military warrior, the secret agent, and the ordinary citizen are faced with a problem whose answer is the ability to fabricate useful firearms from a variety of available products under myriad social, environmental, and psychological conditions.

With the knowledge that they can create a personal firearm from easily obtained raw materials, plus from the knowledge gained through study and experience, free men and women will never be at a loss to defend themselves, whatever the situation (e.g., a fleeing POW, the exigencies of combat and supply, an agent scurrying for a closed, hostile border, or free citizens rising up against harsh police-state dictatorship).

Ever since our early human ancestor first fastened a sharp piece of thin flint to a wood shaft to create that high-tech weapon . . . spear . . . every dictator knows that he who controls weapons controls the populace. A free country permits its populace to own weapons freely only when the government and the citizens have nothing to fear from each other. Unfree countries do not; police states and dictatorships have total weapon control. Stripped of all its sociobabble, the modern issue of gun control is just that simple.

The framers of our federal system of government saw this principle as one of timeless truth: whoever controls weaponry controls the people and the nation. They also recognized that so long as the people had the same access to weapons technology as the government, our nation would remain free of autocratic rule. Our founding fathers would be appalled at the number of attempts by autocrats and their armed supporters to sodomize our Second Amendment to the Constitution.

The major constitutional issue of gun control makes the content and message of this book more important than when the first volumes were published in the 1970s. Today, our basic freedoms (i.e., our treasured Bill of Rights) are in imminent danger. That gun control has become a cultural jihad—a holy war—in the United States is obvious. According to its own leaders, the major objective of Handgun Control, Inc. is the total confiscation of all privately held firearms. Their strategy, and they make no secret of it, is to nibble and chip away at the gun control issue until they achieve total prohibition.

Their argument against universal armament is that we don't live in dangerous colonial times anymore; that the code of the Western frontier is obsolete; that we don't need guns anymore to protect ourselves. If they believe this hogwash, I'd like any one of them to walk down a darkened street early some morning in any of America's nastier 1,000 or so cities, and take along their spouses and kids and pets and some personal valuables. Will anything be left to be found?

No serious, thinking person can possibly believe that once guns are outlawed the entire world will turn peaceful, kind, and gentle and that everything and everyone bad will go away.

First, our world's not kind and gentle, and neither are many of us who live here. Nature is cruel and nasty. You have to be tough to win and to survive. Making artificial rules to ban such technology as weapons does nothing to counter the emotional, biological, genetic, rational/irrational baggage that each of us carries.

My point? Gun prohibition doesn't work. A free people is an armed people. And if you take away free people's personal firearms, as the One World Order Police State that now owns America is trying to do, they will make oth-

Primitive workshops in dozens of villages in northern India turn out homemade guns. (Nicholas Paterson photo, courtesy of Guns Review.*)*

ers. As this book points out, while the method, means, and technology are simple, convenient, and in place, police state dictators provide the motive. Both ancient and current histories are replete with frightening examples.

For example, while on United Nations duty in Cyprus, a unit of the Canadian army recovered a handmade Thompson submachine gun produced by Turkish Cypriots to bypass the arms import embargo. According to Capt. J.M.G. Gagné of the Canadian unit, the weapon was handmade except for the barrel, compensator, and a few minor parts. The two pieces of the receiver were machined out of brass, which did not create problems of reliability. The weapon is one of hundreds that were produced, he reported.

He said the few issue parts were either smuggled in or already existed on the island. The rest of the homemade parts were fabricated in rudimentary workshops that turned out assigned pieces. Pieces were then collected, assembled,

Rare photo taken in 1965 shows underground ordnance repair shop inside a cave in Yemen. Workers produced weapons and repaired or modified the proliferation of Soviet and U.S. small arms in the area. (Courtesy of Time/Life Corp.)

A handmade Turkish Thompson produced in a clandestine series of workshops on Cyprus. Total production of such weapons is estimated at 10,000. (Photo courtesy of Capt. J.M.C Gagné, Royal 22 Regimental Museum, Canada.)

and shipped at night. Captain Gagné said most of the homemade Thompsons went to the Turkish National Guard. He said about 10,000 such weapons were produced.

"The United Nations never succeeded in stopping this

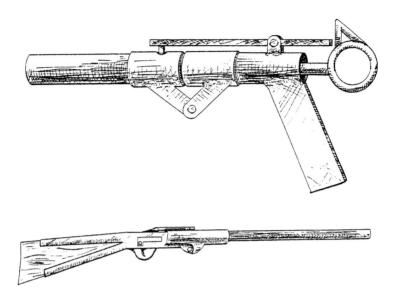

At top, a 12-gauge shotgun produced in Cyprus. As shown below, the trigger releases the cocking handle in the L-slot-type firing mechanism. This weapon can be viewed at the Imperial War Museum, London. (Illustration by Chris Kuhn.)

illicit production," he added. "Once a workshop was identified, the production of that particular piece would then start somewhere else, just as in occupied Europe in World War II."

Every single society or environment that has banned firearms or created stringent control laws has witnessed the practical failure of such prohibition. For example, Japan has very tough laws against private ownership of firearms. To "compensate," Japan became the country where replica or model look-alike guns originated. Many of these so-called unshootable replicas have been modified for actual use.

In 1976, Tokyo police made raids in which 33 replica

Testing a homemade Sten gun during World War II in Norway, a Resistance armorer fires it into a silencing target device in his shop. Note racks of parts of the shop-produced Stens. (Courtesy of Norges Hjemmefrontmuseum.)

Two Haganah members with homemade Sten guns, circa 1947. (Courtesy of Haganah Archives.)

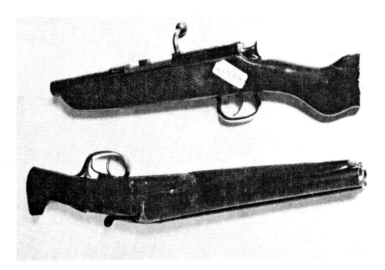

Great Britain's gun laws are quite draconian by American standards, and these two modified weapons show how effective those laws are at stopping illegal weaponry. At top is a shortened rifle, while a sawed-off 12-bore shotgun is shown below. (Official Scotland Yard photo.)

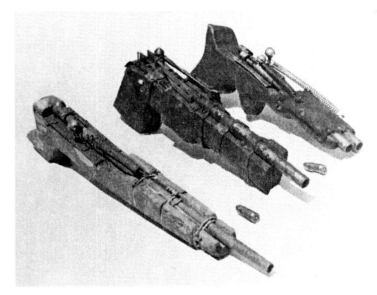

A collection of handmade 9mm zip guns recovered from Mau Mau terrorists during the "troubles," as the British referred to the uprisings in Kenya of the early 1950s. (Courtesy of Donald Steffey.)

British Special Branch officer studies a homemade pistol, with a jungle workshop carbine lying on the mat beneath him. The other items are Mau Mau oathing paraphernalia. Photo taken near Nairobi in 1952. (Courtesy of Africapix.)

guns were seized. Each had been converted to fire live ammunition. One person arrested told police, "We don't have to spend fortunes for black market guns; we can remodel our own guns from the replica toys." By 1979, the practice of conversion had become so widespread that the Japanese government was considering regulations against the replicas. Police had confiscated more than 1,000 converted weapons in 1976, out of the 500,000 replica models manufactured. By 1979, that number had risen to 2,600 conversions out of nearly 1 million models. No further restrictive law was passed, however, because the Japanese admitted the futility of such action.

In 1986, District of Columbia police issued a report that criminals apprehended with what appeared at first to be Uzi submachine guns turned out to be carrying nonfiring replicas. On the other hand, officers were cautioned against assuming that all Uzis would be nonfiring replicas. An ATF report the following year described a

A collection of homemade Mau Mau guns captured from 1952 to 1954. Some triggers are decorative only, pull-release being the method of firing. (Courtesy of Imperial War Museum.)

replica Uzi that had been heavily bolstered and modified for fully automatic firing.

Former ATF Public Affairs Officer Jack Killorin said, "A robber likes intimidation, and the visual sight of an Uzi, real or not, surely is intimidating."

Killorin also noted that the thieves carrying nonfiring replicas thought they would get less jail time than if they had a real gun. They were, of course, mistaken. The feds and most states have laws that treat the implication or

A joint police/army raid scooped up dozens of weapons, including many homemade and modified automatic models, in Northern Ireland late in 1979. The improvised/modified varieties included submachine guns made from machined parts, found materials, and salvaged portions of issue weapons. In addition, several pistols were fashioned in the home gun shops of the IRA. (Courtesy of T.H. Irwin.)

threat of a firearm during commission of a crime as a much more serious offense than actually possessing one.

A replica Uzi converted to full-auto fire was confiscated in Seattle in 1989, while another turned up in Denver in

More IRA goodies recovered by British action, including several homemade submachine guns (center) and two sawed-off shotguns (upper right). (Courtesy of T.H. Irwin.)

This inexpensive Italian-made child's cap gun was converted into a .22-caliber revolver by a 15-year-old who simply drilled out the cylinder and barrel plugs to create a six-shot revolver. It test-fired safely and accurately. (Courtesy of Lombard, Illinois, Police Department.)

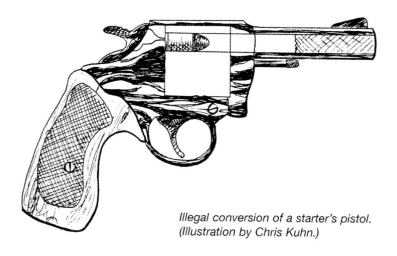

Illegal conversion of a starter's pistol. (Illustration by Chris Kuhn.)

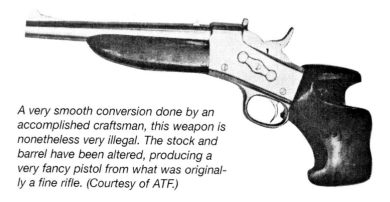

A very smooth conversion done by an accomplished craftsman, this weapon is nonetheless very illegal. The stock and barrel have been altered, producing a very fancy pistol from what was originally a fine rifle. (Courtesy of ATF.)

1990. Each was owned by a street-gang punk. Neither weapon would operate reliably.

As in Japan and the United States, in many firearms-unfriendly countries, citizens are turning to their own ingenuity when restriction requires it. Does this truth then raise the Second Amendment factor in the United States, at least as this book is written? In a constitutional sense, is it legal for you to make a new firearm or to modify an existing one to a new mode of operation or appearance in the United States? The answer depends upon which lawyer you talk to and/or which legal journal you study. The lack of clear

Technically this is a "firearm mode from a shotgun" and is not a true sawed-off shotgun, according to ATF interpretation. In this instance, a 12-gauge shotgun has been cut to an overall length of less than 26 inches, the stock altered, and the barrel cut. (Courtesy of ATF.)

This short-barreled illegal rifle was gunsmithed by some malefactor in New York. (Courtesy of ATF.)

This single-shot .22 rifle was crudely altered by cutting the stock and the barrel. In this case, the weapon is illegal because its overall length is less than 26 inches and it was not originally designed or manufactured in its present condition. (Courtesy of ATF.)

answers to this basic question will make your head spin. There are as many progun legal minds who say that it is absolutely legal as there are lawyers arguing from the other side of the issue that it absolutely is not. My advice is to follow the official rules of the ATF and complete their paperwork that makes you and your creation legal.

If you counter my advice by saying, "This is my free

country and if I want to make or modify a gun that's my God-given free right to do so; I am not a prisoner in my own country," you might be making an ironic point. If so, I hope you have access to a lot of sanitized environment on/in which to hide your illegal weapon(s) and a good lawyer if the feds catch you doing so.

Please follow my advice if you decide to go beyond the purpose of this book, which is for information only. This book is not intended as a how-to manual by any stretch of anyone's imagination.

CHAPTER TWO

Have Another Round

GUNS DO NOT KILL PEOPLE, but bullets do. That old hoary bromide has been rechambered to illustrate that the cartridge and its projectile are the action-dealing parts of a firearm, for without them, there is no kill. If you are in possession of a cartridge, you can build a firearm around it. Or, if you have a firearm but no ammunition, you can certainly fabricate any variety of lethal charge/projectile combinations. The veracity of these facts dispels the tooth-fairy notion that if either guns, ammunition, or both are banned, all firearm-generated crime will go away.

Thus, anyone with either the weapon or the ammunition can create a lethal combination. Seat the projectile in the hardware, then devise a system of aiming and discharge and you've got a weapon. Aiming is done with manual open, telescopic, or electronic sights, while discharge is accomplished either by percussion or electrical initiation.

Whatever the firing system, nothing comes out of the barrel without the cartridge system. These systems may be subclassified as follows:

1. Commercial and military standard ammunition.
2. Reloads of the above, ranging from using modern

presses and dies to reload cartridges to decapping and repriming and shaping the primer cup's match-head paste and then filling the cartridge with a projectile from round stock of a caliber approximating the original.

3. Cartridges tailor-made to fit a particular weapon that is available but for which no issue ammunition is available, e.g., a Viet Cong cartridge made of copper tubing with a plaster piece soldered to the base and a primer seated in the hole.

4. Subloads, which are cartridge cases of smaller diameter than the available weapon that are made to fit by wrapping tightly with paper, plastic tape, or foil to thicken.

5. Overloads, where the barrel is larger than the cartridge chamber, in which case the bullet is pulled and a larger projectile, e.g., marble, ball-bearing, lead slug, etc. is muzzle-loaded onto the cartridge case. This type is common with homemade bolt-action breech-loading weapons.

6. Rimfire cartridges, which are the most common of all types of loads and are preferred. The advantage is that the rim holds the cartridge in place while the striker or hammer hits the primer, which encircles the rim.

Conscription is a way of growing up in the Third World, as this youngster symbolizes, armed with a very lethal homemade zip gun using a spring, a nail, and gas pipe to fire a 7.65mm pistol cartridge. (Courtesy of Max Sink.)

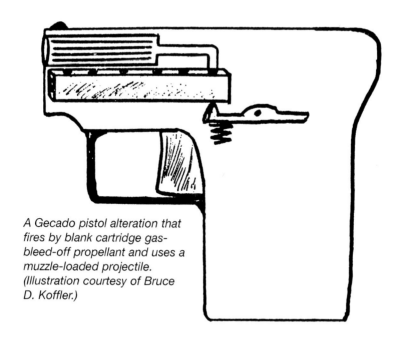

A Gecado pistol alteration that fires by blank cartridge gas-bleed-off propellant and uses a muzzle-loaded projectile. (Illustration courtesy of Bruce D. Koffler.)

Most rimmed cartridges, including those for handguns, rifles, and shotguns, negate considerably the sloppy tolerances of homemade breeches and the attendant problems of cartridge seating and head spacing. The extraction is easier, and the rim adds another zone of strength to the cartridge with a greater margin of safety to the shooter.

Because a great many improvised firearms are simple muskets with smooth, large bores, like the blunderbuss of old, bullet options are varied (e.g., hunks of glass; pebbles; slugs; sinkers; nail ends; split washers; hardened candle wax, which creates nasty wounds up close; ball-bearings; split-shot sinkers; and a few other junk-pile goodies, the dirtier the better).

Packing phonograph needles or headless finishing nails is a way of improvising flechette charges in shotshells. More esoteric charges, such as aluminum paste and Silly Putty, have been used for special loads.

For muskets loaded with lead balls, accuracy can be improved by drilling a hole in the base of the ball and inserting a short wire tail and securing it with melt heat or by peening it into place. In flight, the tail will project behind the ball, stabilizing it.

Glass bullets, such as beads and marbles, are lethal due to their fragmentary propensities. Also, it is impossible to trace such bullets within a victim's body except with X rays. Plastic, although lighter, has the same effect.

All ballistics experts suggest that hollowpoint ammo should be used in preference to military ball. When this is not possible, the ball ammo should be deformed to induce aerodynamic instability for short-range buzz-saw and irregular in-flight tumbling, i.e., keyhole effect, in wounds.

This, of course, brings to mind the "dumdum" bullet of infamy and much fiction. A true dumdum bullet requires that at least one deep slot be cut into the tip to induce flattening of the projectile after it enters flesh. Cutting an X slot produces fragmentation of the bullet, creating a very nasty wound.

Filling a standard hollowpoint bullet with strychnine or cyanide has been done, as has adding a common BB shot to help split the bullet as it enters the body. The mercurial droplet a la the Jackal is another touch, as is the old Sicilian trick of packing the bullet cavity with garlic or fecal matter.

I interviewed one former terrorist, now in prison, who told of improvising explosive ammunition using a standard rifled slug or shotshell lead ball. He placed a .22-caliber blank cartridge in a well drilled in the nose of the host projectile. This causes deformation of the slug when the blank detonates upon hitting a bone.

"This results in some really spectacular and very ghastly body damage," he added.

An American Special Forces veteran of Vietnam, El Salvador, and Desert Storm told me of creating "experimental" rounds by placing large pistol primers in the wells of issue hollowpoint ammunition. He said he saw the idea field-tested in several theaters in .38 Special, 9mm, and .45 ACP.

"I never saw this but was told about a couple of American snipers who did the same trick with some 7.62mm rounds they had hot-loaded over in Southeast Asia . . . bet that really blew away those little suckers," he related.

Poisoned, dumdum, explosive, dart, glass, hardware washer-stacks, and ball-bearing loads are all improvised payloads for improvised/modified firearms. They have a common bond among them, namely their capacity to produce a ghastly wound, even with a marginal hit. Their use range varies from a few inches to a couple of yards. Long-range accuracy is not a consideration; short-range lethality is. When there is only one shot, one chance, it has to be good, and anything that will give that lethal edge must be used.

This was a lesson taught well by guerrilla soldiers during the Vietnam conflict. USAF S. Sgt. P.D. Long recounted a story in which Viet Cong gunsmiths converted ammunition to fire in an existing pistol. "In 1972, while I was in Southeast Asia working for an explosive ordnance disposal team, an army major handed me a Chinese copy of the Soviet Tokarev pistol," he said. "The proper bottle-necked 7.63mm round was extremely rare in this area, so the Viet Cong could not get that round. To use the piece, they made their own ammunition.

"First, they took a 9mm Luger round, pulled the bullet, but saved the powder. Then, they took a .32 ACP case, cut the head off, and slipped the case and bullet into the 9mm case. A slight crimp at the joint held the assembly together . . . The major assured me the combination worked fine."

At this point, Staff Sergeant Long added his most important point: "Mere difficulty of obtaining something—legal or illegal—has never stopped a determined man from obtaining it, or the materials for producing it."

3 CHAPTER THREE

Pipes, Plumbers, and Other Spare Parts

CONSTRUCTION OF TRULY IMPROVISED firearms relies mostly on the available cartridge. The ubiquitous .22 cartridge is the favored caliber. Its availability is doubtless the major reason, but there are other factors worth considering.

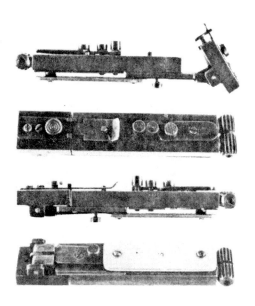

This homemade firearm is being produced and sold along the Mexican/U.S. border for less than $35. It measures 5 1/4" x 1 1/2" and weighs just 9 ounces. It fires two .22 LR cartridges, has rifled barrels, extractors, plus a firing and cocking device for one or both barrels. The sample submitted to the FBI was rated as excellent in workmanship. (Courtesy of Chula Vista, California, Police Department.)

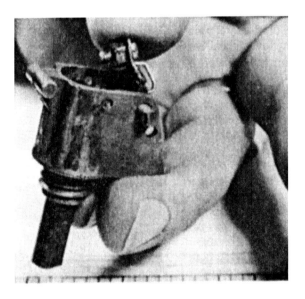

This unusual .22-caliber firing device was made by taking the detonator top from an inert grenade, drilling out the primer area, and inserting a .22 round. The firing pin is spring-loaded. (Courtesy of Oceanside, California, Police Department.)

PIPES, PLUMBERS, AND OTHER SPARE PARTS • 33

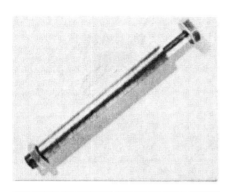

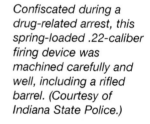

Confiscated during a drug-related arrest, this spring-loaded .22-caliber firing device was machined carefully and well, including a rifled barrel. (Courtesy of Indiana State Police.)

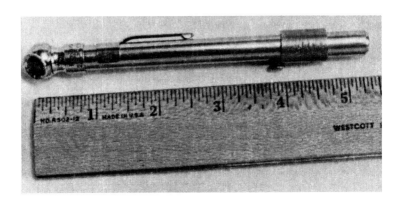

A very well-disguised .22 firing device was recovered from an outlaw biker during a drug-related arrest. Built into a tire gauge, this single-shot weapon shows excellent machining and design. It fired safely and accurately at short ranges. (Courtesy of FBI.)

The .22-caliber round is generally very safe to the improvised gun firer. Even in weapons where a breechface is nonexistent, the thickness of the cartridge base is such that it can withstand the firing pressure virtually unsupported. Cases often bulge and sometimes burst. Yet, even when they do, the .22 cases simply peel back rather than fragment, presenting little danger to the shooter.

The barrel for a .22 zip gun usually is a section of car antenna flared to fit and then taped to a wooden grip. Other .22-caliber barrel expedients include grease-nipple rods, threaded lamp-cord rod, and bar stock drilled with a 15/64 bit. The firing pin is often a nail or a sharpened small bolt driven by a spring or rubber band.

Accuracy tests are meaningless, as these .22 weapons are for point-blank distances only. Penetration varies proportionately to the amount of breech support at the base of the cartridge. With well-made zip guns, the results are the same as those obtainable from pistols; most are not well-made.

The user-friendly, target-harming tendency of these projectiles to keyhole often makes formal penetration testing academic. However, penetration of at least a 1-inch pine board is possible even with the crudest .22-caliber zip gun firing a keyholed bullet.

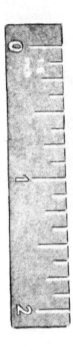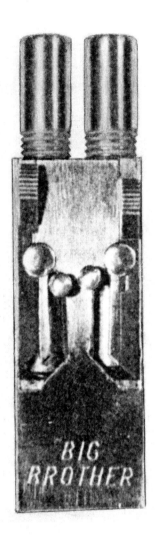

This clever handmade antique firing device is a double-barreled .22-caliber weapon used for close-quarters mayhem. (Courtesy of Alex Foley.)

One police officer said, "Those keyhole .22 slugs are dangerous as hell, even out of a street-crude zip gun. I'd sooner face a drunk with a standard piece than some punk hoodlum with a keyholer."

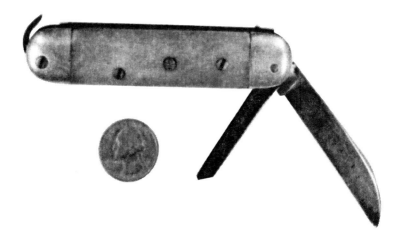

An unusually well-disguised knife gun, the German Resl is a .22-caliber pistol with its barrel parallel to the blade and its trigger the serrated "screw" in the middle of the knife grip. (Courtesy of David H. Fink.)

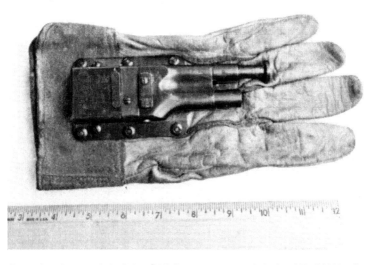

A production model of the ONI fist gun as used during World War II. (Courtesy of Maj. Richard Keogh.)

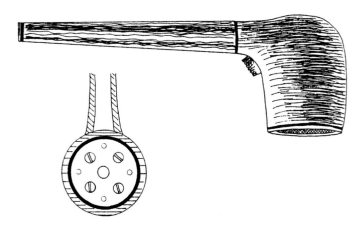

Home Guard .22-caliber pipe pistol. (Illustration by Chris Kuhn.)

This one-shot zip gun was made in Los Angeles from a child's cap gun. The swastika was added by the maker. (Courtesy of Los Angeles Police Department.)

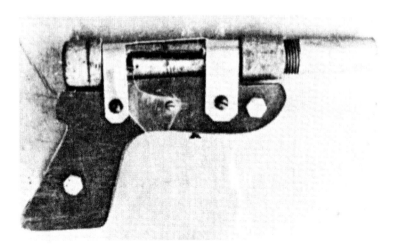

A plumber's helper—a zip gun made from pipe, Plexiglass, tin, and bits of wood. According to the LAPD's lab people, it shoots well. (Courtesy of Los Angeles Police Department.)

The U.S. Army made extensive tests on the keyholing effect of its .223 ammunition used in the M16 rifle. In simple terms, it found the power and energy of the keyholing round transfers very quickly to the target. If that target happens to be human, the result can be a messy, frightful hole.

Police officers facing an individual armed with a .22-caliber zip gun are urged to use extreme caution, as a lot of these items are pull-release; thus, dead-finger firing is a distinct possibility. Or, in a hostage situation, the zip gun could be pointed at the hostage and the striker drawn back. Any attempt to disable the kidnapper should be discouraged until the kidnapper is convinced to surrender or is somehow lulled into easing the striker back to the base of the cartridge.

Moving from .22 weapons to the 12-gauge shotgun is the other end of the spectrum. As this cartridge also has almost universal dissemination, the 12-gauge is used worldwide in homemade weapons. This cartridge is ideally suited for improvised use, as its dimensions are such that it will easily chamber into three-quarter-inch plumbing pipe. Most improvised 12-gauge shotguns utilize this pipe for

Pipes, Plumbers, and Other Spare Parts • 39

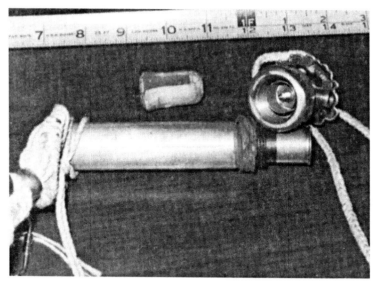

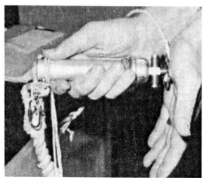

A Seattle police officer stopped a suspicious man carrying a small piece of plumber's pipe. Luckily, the pipe was not used, as it turned out to be a loaded, homemade 12-gauge shotgun. The weapon is fired by slamming the hand into the shell cap. (Courtesy of Seattle, Washington, Police Department.)

the shotgun barrel. Apart from banning plumbing pipe, there is little hope of the antigun movement or, for that matter, any police state getting the upper hand on the manufacture of these illicit shotguns.

As any combat soldier or serious shooter will tell you, a shotgun is far more lethal at close range or in most environments than a submachine gun. Perhaps the most lethal close-range weapon would be the full-auto shotgun. For now, let's consider its poor-side-of-the-tracks neighbor, the plain-folks' homemade scatter gun. Many are short-barreled and, due to the use of the three-quarter-inch pipe with no taper, have open bores (i.e., no

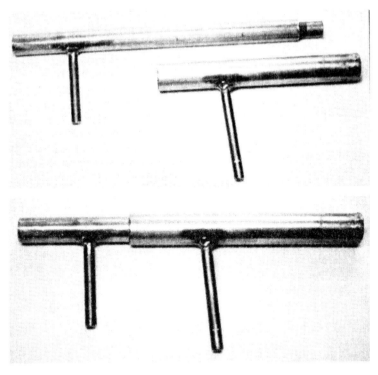

This 12-gauge "slap fire" shotgun was constructed of steel tubing by a teenager and is quite mechanically sound. (Courtesy of Brunswick, Ohio, Police Department.)

choke). This makes them pattern wider, a major benefit in a single-shot weapon.

Plumbing pipe endcaps become breechfaces, T-joints are grip assemblies, and threaded joints become receivers and stocks. There is little worry about safety factors because the strength of galvanized plumbing parts is considerable. The pipes are thick and will withstand the pressures of the 12-gauge shotshell easily. But there are other problems.

Extraction of fired casings, a problem common to most homemade weapons, is always a hassle. The expansion of the cartridge case on firing causes it to swell and sometimes crack in the breech. Extraction is normally accomplished by ramrodding the empty casing out of the weapon or having a portion of the breech filed away so

Don Walsh, instructor at the Correctional Staff College Museum in Kingston, Ontario, Canada, examines the 12-gauge pipe gun made by an inmate in the Collins Bay Prison for a breakout attempt. (Courtesy of M.E. Millar.)

as to permit a fingernail or knife blade to get hold of the cartridge rim and extract it. Of course, if it swells abnormally, the firer is out of action.

Between the extremes of the 12-gauge shotgun and the .22-caliber zip gun, there is a great deal of middle ground in the so-called raw materials people use to turn out improvised firearms. For instance, some of the firearms I've seen have no barrels as such. Naturally, any weapon can fire its deadly load without a barrel (e.g., for extremely close work, a revolver barrel may be cut off totally or unscrewed and left at home). Upon seeing such a weapon, a police officer noted, "Something like this would surely give the forensic investigator some true pause for concern!"

Some penguns and subcaliber devices either have no real barrel, or the barrel is more of a gutter guide than a close-fitting bore. The British ground spike pistol, although technically a booby trap and alarm, has been used as a handheld firearm, and discharged when pressed against a victim.

Leslie Smith, a noted forensic scientist, has done extensive research on modified and improvised firearms. His case histories are legion, as any student of this science knows. He reports one instance where a taxi driver was killed by a man carrying a stockless .22 autoloading rifle with the barrel cut down to about 5 inches. The fast-firing weapon was simply a trigger group, receiver, magazine, and stump of a barrel.

A similar weapon, known popularly as a "boot gun," was recovered by a Florida sheriff after a youthful burglar wounded the lawman with the gun. It was a .22-caliber single-shot rifle with the stock and trigger guard removed and the barrel cut to about 2 inches. The overall length was 7 inches.

According to Smith, "This type of gun is often carried in a boot or a side pants pocket. They often turn up in gang fights, sometimes with a modified pistol grip."

Former police lieutenant Dr. Joseph Bogan, now an esteemed criminologist, has done a great deal of research into juvenile gang warfare and reported, "For some reason the punks really go in for the sawed-off weapons, which is a bit ironic, considering how overrated they are."

He's right. Most sawed-off shotguns make great assassination weapons when the target is in a telephone booth or a narrow hallway. But beyond that, well, perhaps an anecdote will scratch away the glamorous facade of sawed-off weapons, exposing the tarnish of inaccuracy.

A police report noted that a young man accidentally shot his sister-in-law while she was using a rustic outhouse at the family camp. The young man was on a backyard range testing his freshly sawed-off .22 rifle.

According to the report, "He was firing at a target some 50 feet downrange and about 15 or so feet from the outbuilding. His shot missed the primary target by some 12 feet and hit his sister-in-law, who was in the outbuilding. Both the victim and the police are totally satisfied the shooting was accidental. The man was charged with having an illegal weapon, though."

No sawed-off weapon is intended for ranges beyond 10 feet or so. They are arm's-length killing weapons, effective in small rooms, alleys, or into car interiors. The sawed-off firearm's only serious use is where concealability or point-blank murder is the major factor. Period. Except for those two prerequisites, the sawed-off belongs to television and pulp fiction.

U.S. Customs officials always turn up interesting weapons during their searches. Indeed, weapon smuggling has become so creative since the new security regulations at airports that one Customs official remarked, "Some of the stuff we get makes James Bond's exotic arsenal look like so many boutonniere water squirters."

One such device was dubbed "Secret Sam's Pipe Shooter" by agents. It was an ordinary tobacco pipe, capable of being smoked and of firing a .22-caliber short cartridge from the base of the bowl. The user has only to bite on a hidden pressure trigger in the stem to fire the weapon.

At New York's La Guardia Airport, a young lady was arrested for trying to carry aboard a small, German-made autoloading pistol hidden in a radio. Meanwhile, at JFK agents seized a pseudo-blind man's cane that contained a firing device. Motorcycle gang members often use these hidden modified weapons, too. For example, a young man

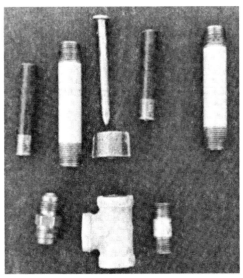

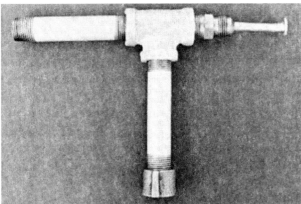

Muncie, Indiana, police confiscated this .410-gauge pistol made entirely of hardware items. A large spike nail fires the plumber's pistol. Interestingly, the grip uses a pipe cap and provides storage for an extra shotgun shell. (Photo courtesy of Muncie, Indiana, Police Department.)

was arrested in Canada and charged with being an armorer for a local motorcycle gang in an urban area of Ontario. Police found a cut-down, semiautomatic .22 rifle with a sawed-off barrel in the man's car.

In New York state, police recovered several modified firearms from motorcycle gangsters, including a sawed-off

This Universal .30-caliber pistol was created from an M1 carbine, using a 30-round magazine. It was evaluated and tested by the Montreal Police Department's Technical Section. (Courtesy of Jack Krcma.)

shotgun, a shotshell pistol homemade from a .410 bolt-action shotgun, and an unsuccessful attempt to convert an autoloading pistol into a full-auto weapon. In Pennsylvania, state police arrested motorcycle riders following violence near Pittsburgh. Several modified firearms were found, including tear-gas guns converted to fire .32-caliber ammunition. Later investigation of their farmhouse turned up two M1 carbines illegally converted to full automatic capability.

Other unusual items confiscated from bikers in Michigan included two .22 pistols created from tire pressure gauges and a bicycle pump rigged with a spring and a 12-gauge shotgun shell. Nobody with the Michigan State Police wanted to test-fire any of the three, so they rigged them into a vise and discharged them by remote control. All fired safely.

According to the Customs agents on the Mexican border, motorcycle gangsters have revamped their vehicles to hide all sorts of weapons and ammunition (e.g., shotshell-firing handlebars, gas tanks modified to carry sawed-off shotguns, and shock absorbers converted to fire a shotshell).

Former federal agent Dan Walsh mused, "What ever happened to the kindly old days when bikers used to smuggle only pot in their gas tanks?"

While hardly in the homemade category, the exoteric weapons turned out by CIA armorers could be classed among the unusual samples presented in earlier volumes of

Can you dig it? It's a shotgun shovel, an unusual design by a local biker made from a shovel handle and galvanized pipe. The barrel is threaded pipe while the firing mechanism is a spring-loaded metal rod. (Courtesy of David H. Fink.)

improvised and modified firearms books. Certainly, much of their gimmick ordnance is merely modification of existing weapons or designs.

Sadly, though, when news of these weapons is leaked,

A machinist built this gun-in-a Pepsi-can for Mr. Fink's collection. It's a well-disguised .50-caliber percussion pistol built into the can. (Courtesy of David H. Fink.)

the media concentrate on the hardware, on the exotic gimmicks shown them, like so many small boys peering around a toy store at Christmastime. They never get to the inhumane side of the story.

When the U.S. Senate's Select Committee on Intelligence (a.k.a. the Church Committee) got around to the stage-managed presentation of the Agency itself, the CIA's propmaster had arranged a suitable display for the senators to cluck over.

As *Quill*, a respected journalism magazine, reported in 1979:

> The Committee was treated to a dazzling display of James Bond devices. Developed under an eighteen-

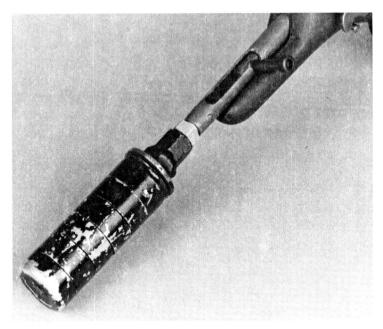

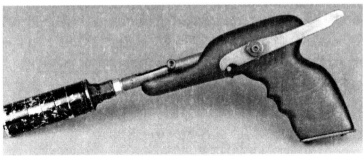

This homemade single-shot .32 pistol is equipped with a suppressor made from a modified garden tractor muffler. The weapon is fired by the spring-release trigger bar. (Courtesy of ATF.)

month, $3 million covert research program called "Project Naomi," the arsenal included dart guns . . . light bulbs and automobile engine parts that emit poison gas . . . bizarre weaponry . . . umbrella guns, silent guns . . . Somber-faced legislators were duly photographed pointing dart guns or looking with suitable dread at a bottle of shellfish poison.

Pipes, Plumbers, and Other Spare Parts • 49

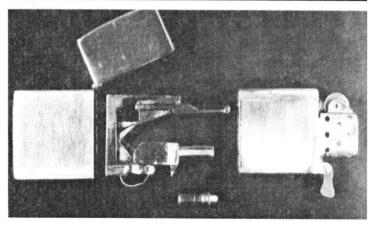

The smoothly turned-out firearm within a cigarette lighter is a real threat to law enforcement people. Recovered by the Aurora, Colorado, police, this weapon fires a .22 LR cartridge from a hidden trigger about a quarter inch above the flame guard. The three photos show the clever design and machining of this weapon. (Courtesy of Aurora, Colorado, Police Department.)

Attaching flashlights to guns is hardly news, but incorporating a firearm into a flashlight really is rare. This heavy-duty lethal light housed two .38 Special rounds and was named "Protecto Lite." It was commercially produced but ran afoul of federal laws and was discontinued. (Courtesy of David H. Fink.)

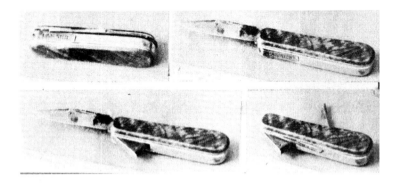

This single-blade pocket knife turns into a double-edged threat when the .22-caliber facility is uncovered. The weapon is loaded, and the barrel is closed back into the knife handle. The cocking/firing lever is then opened out and squeezed. Bang. These types of devices are quite common—the sample is uncommonly sophisticated. (Courtesy FBI, Law Enforcement Bulletin.)

PIPES, PLUMBERS, AND OTHER SPARE PARTS • 51

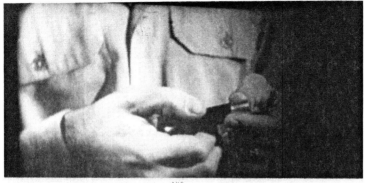

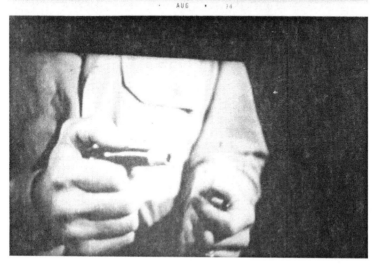

Three views of Alan Ladd in the wartime propaganda film OSS, showing the cinemasmith's view of the pipe pistol. At top, Ladd is "smoking" an innocent pipe; in the center frame, he's preparing to fire; at bottom, he fires. (Courtesy of Peter Senich, American Historical Research.)

A German pipe gun in .22 caliber (top) and an American-produced pipe gun, also in .22 rimfire (bottom). The American gun is named "Sweetheart." Both weapons date from the middle 1930s. (Courtesy of David H. Fink.)

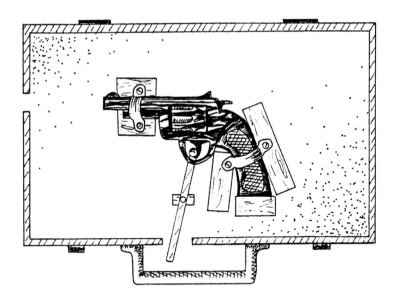

Attache case altered to fire a pistol. (Illustration by Chris Kuhn.)

Lost in the thrill of newsmen turned little boys looking at toy gadgets, reporters again failed to ask "why" and "against whom?"

As Allen Dulles, the old spymaster, used to yawn to

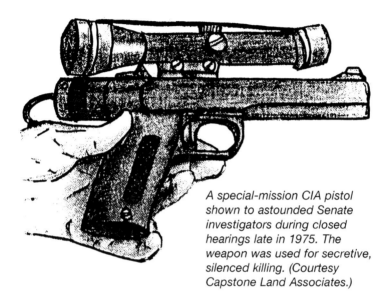

A special-mission CIA pistol shown to astounded Senate investigators during closed hearings late in 1975. The weapon was used for secretive, silenced killing. (Courtesy Capstone Land Associates.)

aides before venturing into congressional hearing rooms, "I'll just tell them a few war stories." And that, certainly, is feud for fought.

4 CHAPTER FOUR

Familial Feud

NOT ALL OF THE FIREARMS THAT KILLED Soviet troops in Afghanistan in the 1980s came from sanitized CIA armories or were stolen from the Soviet invaders. Many were manufactured in cottage industry plants in or near Darra, the arsenal for the Pakistani Northwest Frontier and major supplier to eastern Afghani rebels.

For more than 100 years, Darra has supplied copies of all major military weapons to Pathan tribesmen and others who roam the inhospitable mountains between the two nations. While the factories are in the hills, the gun shops are along Darra's dusty streets, interspersed with shops selling teas, spices, marijuana, or hashish . . . all quite openly.

The locally made gun copies are both proliferate and proficient, e.g., well-made and fully operational Sten submachine guns, British .303 Lee Enfield MkIV rifles, Webley revolvers, and ubiquitous Soviet AK-47 assault rifles. Producing enough firearms isn't the problem; getting ammunition is.

In 1982, a rebel officer told Michael Fathers of Reuters: "Our main problem is a shortage of ammunition. Someone needs to open a production plant for that. We do not have

Oldsters and children work side by side in Darra's famed cottage industry, turning out copies of all sorts of small arms in primitive plants such as this. (Courtesy of Helms Tool & Die, Ltd.)

enough ammunition. If we had that we would have taken Kabul [the Afghan capital] years ago."

Within two years, that plea was answered. One ambitious gun maker in the region had produced six fast-firing antiaircraft weapons, which he sold to the rebels for $2,000 each; compare that to the $15,000 to $25,000 that the United States paid for its .50-caliber version, or the Soviets for their 12.7mm model.

According to Stuart Auerback of the *Washington Post*, there are 4,000 arms makers in the region turning out everything from copies of Webley and Mauser pistols to SAM missiles and 20mm antiaircraft guns. One maker, M. Ishoq and Brothers, has been turning out duplicate weapons, right down to the "Made in Germany" or "Made in the USA" stamps, for nearly 70 years. They produce M16 rifles, Sten

A gun merchant in Darra, on the Pakistani frontier, sells handmade copies of rifles, shotguns, pistols, and submachine guns. The roar of gunfire rules this wild land. (Courtesy of Helms Tool & Die, Ltd.)

guns, Webley pistols, Mauser pistols and rifles, and the H&K G3 assault rifle. Prices range from about $100 for a pistol to $3,000 for a heavy machine gun. Producing scarce ammo brings about 30 cents per 9mm round to $3 for a .50-caliber round.

As Catherine Lamour and Michel Lamberti point out in their book *The Second Opium War*, in Pakistan weapon counterfeiting is truly a national industry among the independent mountain people of the Northwest Frontier Province. I have seen examples of high-quality copies of the M16A1, as well as the newest Czech, Polish, Spanish, and Italian submachine guns, and even top-notch mortars and light artillery—exact copies, down to the counterfeited proof marks.

However, the quaint image of native craftsmen bending over crude hand tools is a bit too much *National*

58 • ZIPS, PIPES, AND PENS

In the wild Pakistani frontier, a Pathan fires his Sten copy. Craftsmen produced this quality copy of the British Sten in their own workshops. (Courtesy of Helms Tool & Die, Ltd.)

Geographic for today's reality. While a few weapons are made that way and some components are still hand-forged, most of the shops now use electrically powered lathes and other power tools for a faster run at mass production, fed by profits from the war across the border in Afghanistan.

Noting this, Italian journalist Ren Savoldi commented:

"Perhaps the mountain arms sellers saw what was coming to the Middle East back in 1979, because they started to modernize and mass produce for the coming troubles. A lot of weapons and rupees changed hands that year." He was correct. As the Soviet invasion of Afghanistan ushered in 1980, the illegal arms markets in neighboring Pakistan became a real seller's paradise. There were some unofficial reports that while the Pakistani government waited for the United States to lift its arms embargo that year, its military entered into purchase agreements with the domestic weapons counterfeiters.

According to Savoldi, the quality of these copied weapons from Pakistan varied greatly, depending upon the shop and the metal used. Some used cheap railway iron while others used hardened steel alloy imported from Western Europe. "You can buy cheap merchandise or you can buy expensive quality," he wrote. "It's just like buying ordnance anywhere—you get what you pay for."

Neither the government of Pakistan or Afghanistan has any control or force in the area. And, as one dealer told Savoldi: "Neither did the damned British when they tried in the last century."

According to former intelligence officer and ordnance salesman Jack Krcma, "This area outdoes the old Wild West at its worst. It makes a Sicilian vendetta look like a nursery school picnic. This is the Northwestern Territories, where East meets West in the Hindu Kush—the Khyber Pass."

These family battles have been going on for generations, as blood oaths are sworn and a brother's murder must be avenged, creating a deadly, vicious circle of vendetta. Gunfire pierces the Khyber night and thunders from dusk to daybreak. Then the tally is noted, and the balance is brought forward into the next day. The homemade firearms are quite functional and, considering what the locals have to work with, well made.

One store specializes in pistol versions of the .303 Enfield. Firing a .303 pistol is a major shock. After the sheet of flame and the roar of sound, it's a surprise to see your shaking hand is still attached to the end of your arm. The recoil is strong but not excessive, and the .303 round is a close-range killer. The .303 pistol sells for $150, Krcma says.

These hardy folks of the Northwest Frontier were never really subjugated by the British. The natives kept the regular British army pinned down for nearly a hundred years, such that this colonial frontier was regarded as a finishing school to test the mettle of the Empire's soldiers and officers at that time.

At first, the hill tribesmen were armed with matchlocks and flintlocks, which they had copied and built from Persian and Afghan weapons. They pitted these weapons against British pack cannons, machine guns, and bolt-action rifles, battling the regulars to a stalemate.

When they acquired battlefield samples of modern repeating rifles, like Martinis and Enfields, they copied them, too. Scrap iron, charcoal forges, hand lathes, hammers, and files, coupled with the tribesmen's keen eye for detail and mechanical ability, created a cottage industry for building firearms.

Ted Kale, a freelance journalist who visited several village gunsmiths in the late 1960s, reported seeing craftsmen who turned out expert copies of existing weapons. He was in awe of what these supposedly simple men did with their crude tools.

"They are truly artists and were quite proud of their handiwork," he said. "Although they did hint rather darkly at the use for these weapons, I never saw anyone actually shot, but I heard pops at night and saw one lad who was brought in wounded early the last morning I was in the village."

Kale, who had served in the area during World War II, told me that even the bond of military service and the universal camaraderie of combat did not totally melt the strong suspicion and distrust he felt.

"It was rather like feeling that I had a constant bull's-eye painted on the back of my jacket, you know," he explained.

It wasn't that the fierce tribesmen hate only the

British; they fight everyone. The neighboring tribes and armed caravans are often targets for plunder. In addition, they have a history of arguing among themselves, and there are feuds that continue to this day.

The Khyber copies are not just curios sold to a few tourists, journalists, or tribal hunters; most are, in fact, deadly serious weapons made for murder. That sordid fact is universally found further to the east as well.

5 CHAPTER FIVE

Asian Genesis

YOU'VE GOT TO GO BACK AT LEAST to the year Babe Ruth hit 60 home runs to chart the beginning of the Communist revolution in Vietnam. Emerging late in 1927, the Communist group there was known as the VNQDD. Its credo was national liberation—radical thought for that time and place.

In 1930, the VNQDD caused an armed rebellion called the Yen Bay Mutiny, which precipitated the arrest of 10,000 Vietnamese and the deaths of at least that number. The revolt was quashed by the French, a loss which sent Ho Chi Minh, the founder of the Communist group, into exile in Europe. Despite the defeat, the uprisings gave way to a new popular struggle among the progressive Vietnamese.

The growth of the religious sects like Cao Dai in the 1930s, plus the slide of the Vietnamese Communists into a political and armed force with considerable clout, caused French repression also. The French treated these threats, real and otherwise, seriously, and it wasn't until the Vichy regime was set up during World War II that anything actually challenged their rule in Indochina.

During World War II, Vietnam was quickly occupied, and the country was opened to the Japanese conquerors, although the French were allowed to administer it under Japanese supervision. That symbolism was quickly noted in the image-aware Far East. The white man had been defeated and humiliated by the Oriental. No longer would an Oriental bow before a white master. Japan had bested France; this was a major event in the history of Indochina, and in Asia as a whole.

Despite this setback, U.S. Office of Strategic Services (OSS) and British Special Operations Executive (SOE) forces worked well with native troops to defy the Japanese and their Vichy puppets militarily during World War II. Even if the example of guerrilla warfare was not set at this time, it was surely refined and taught to a generation of Vietnamese by these commando experts from England and the United States.

"The Communist troops were among the best we had. They were disciplined, responded well to our training and orders," recalls former SOE Cpl. Arthur Willis from his experience working with natives in World War II Indochina. "About the only thing I found at fault with them was their politics."

Given the centuries of harsh colonial rule, it's little wonder that socialism, Communism, populism, or anything seeming like self-rule politics appealed to the more literate and outspoken Vietnamese. Consider that the Vietnamese leader most widely admired by the OSS men in the field was Ho Chi Minh, whose major desire, looking beyond Communism, was nationalism.

As the famed reporter and Vietnam scholar Bernard Fall wrote:

> Perhaps the major reason Ho turned to the Communist bloc was because the tight-lipped, tight-collared, United States State Department had its stupid blinders on . . . Had they listened to a few of the boys who sweated in the jungles as junior OSS officers, this mess (the Vietnam War) might not be happening now.

It's not surprising to intelligent historians that when the Allies won World War II, Charles DeGaulle sent personal messengers to the remainder of the French collaborationist bureaucrats who'd served the Japanese in Vietnam, telling them to surrender with the enemy. Yet, according to nationalistic Vietnamese, instead of surrendering, the Japanese seized total control of Vietnam in June 1945 and held on until late September, weeks after the nominal VJ surrender date.

Officially, the United States was against continued French rule in Vietnam for a number of reasons. The French colonials wanted business as usual. For their own colonial reasons, the British sided with the French. The weak-kneed American State Department smiled its way to a bad decision, and the corruption of colonialism was allowed to resume, which fueled local discontent.

Although France held paper title to the country, with United States and British agreement, armed resistance began again across Indochina. A force known as the Viet Minh, both Communist and non-Communist, began to push its liberation movement. Abandoned by his wartime British and American allies, Ho Chi Minh turned to the USSR and China for support.

As Bernard Fall wrote, Americans are not so quick to scream "RED" about Communists helping American GIs in France, Italy, Poland, and even in Indochina prior to 1946.

The freedom movement was sporadic. In addition to the political elements, the religious Cao Dai and the Wa Ho sects were involved, as were the Binh Xuyen bandits, better known as the Vietnamese Mafia. Had these diverse elements ever gotten totally together, the history of the Vietnam War surely would have been different.

Used in the fighting were hoards of World War I and World War II weapons, plus the products of hundreds of workshops and individual technology. The Communists were holed up in Viet Bac along the then Nationalist Chinese border, where they manufactured copies of standard pistols, rifles, shotguns, ammunition, and grenades in their own shops.

Both the Viet Minh and the Viet Cong made extensive use of homemade, copied, and improvised firearms. In this VJ Day picture taken in Hanoi in 1945, several of the men in the foreground are brandishing homemade pistols. (Courtesy of Frank C. Brown.)

Early guerrilla fighters in Vietnam used a variety of available weapons, especially homemade models . . . including the Thompson in the second row. (J.R. Angolia photo, courtesy of Frank C. Brown.)

Saigon in the early 1950s: a number of improvised MAT 49 and Sten submachine guns. (J.P. Gillet photo, courtesy of Frank C. Brown.)

French intelligence sources estimated that in 1946, the Viet Minh manufactured close to 30,000 pistols, rifles, submachine guns, and shotguns. They produced many of these weapons under the guise of doing work at Trans Indochina Railway repair depots. They also liberated factories that manufactured matches, pyrotechnics, and small steel and iron works, then set to work improvising weaponry from this base. The weapons left much to be desired in terms of safety, range, and cosmetic appearance, but they were satisfactory for the close-in fighting of guerrilla warfare. By 1948, the number of weapons the Viet Minh had on hand from surrendered Japanese and French stocks had doubled to an estimated 50,000, and of these, the majority were produced locally.

Other weapons used were those captured or purchased from arms dealers in Southeast Asia or supplied by paradrop during World War II, plus the locally produced homemade weapons, which were a considerable part of the Viet Minh supply. These homemade guns would assume an even greater importance when it came time to resupply or

Chinese Communist troops were often armed with arsenal-turned copies of Thompson submachine guns, produced without license. Armed with the bogus Thompsons, troops march against the Japanese in 1937. (Courtesy of Frank C. Brown.)

Rare photo shows Chinese sailors in the 1930s armed with both antique cannons and vintage local copies of Mauser pistols aboard a trading junk of that era. (Courtesy of Frank C. Brown.)

Marine PFC D.S. Loosemore with homemade ChiCom-type carbine taken from a Vietcong guerrilla, spring of 1966.

outfit new irregulars once the modern World War II weapons had given out.

Over the years, I talked at length with the late Edward Lansdale for several projects I was involved with concerning terrorism and psyops warfare. Often our conversation turned to his early days in Vietnam, and he once told me about visiting the guerrilla camps of the Lien Minh. General Lansdale said he was amazed at the pristine condition of the small semiautomatic pistol carried by his host, Trinh Minh Thé.

"He told me it was a smaller hybrid copy of the Soviet Tokarev, and I asked him where he got it," General Lansdale recalled. "He told me it was made for him in the same place that his people got most of the weapons they did not capture in battle. He took me to their armory . . . I saw everything from M1 rifles to M1 carbines to Soviet PPSH submachine guns to several .50-caliber M2 heavy machine guns. Although some were GI issue, a number of the weapons showed variations and a bit rougher workmanship. They were homemade," he added.

Trinh Minh Thé told General Lansdale that the Lien Minh made all of the copy weapons right there in camp.

Captured Vietcong weapons included myriad homemade, fabricated, and modified firearms, along with the usual U.S. weapons captured in fighting or bought on the black market. These were gathered during Operation Chinook, north of Hue in 1967. (Courtesy of L.Cpl. B.L. Axelrod.)

He showed the American officer, at the time working for CIA, a very sophisticated jungle workshop, complete with forges, drills, lathes, and a variety of modern machine tools powered by diesel generators. Most of the equipment was German, while the shop foreman and senior skilled machinists were Chinese, driven out of their homeland by the Communists in 1949. Even so, it was tough to figure whom they hated more, the French or the local Communists, Lansdale said. "Their homemade weapon which

Frank Brown with an original, locally produced M203—an improvised combination of the M16 and the M79. The idea originated in the mid-1960s in Vietnam. (Courtesy of Frank C. Brown.)

impressed me the most was an exact copy of our big Browning M2. Looking at an issue gun alongside their copy, it was impossible to tell them apart," he told me.

He said the Vietnamese got parts from government arsenals and other raw materials, like steel, by foraging the countryside. He said they showed him a large stockpile of steel rails from a nearby railroad track that had been dismantled.

"'It was for French growers, so we thought it best we convert their train track into our arsenal,' Trinh Minh Thé told me," added General Lansdale.

By this time, though, it was obvious that North Vietnam was arming itself and its insurgents in the south with weaponry from China and the USSR, General Lansdale told me. Yet, while the North Vietnamese stock of modern weapons was finite, their homemades were limited only by scrap metal and ammo until China came to their aid. With the hope of taking pressure off Korea, she committed herself as a base area for the Viet Minh and a source for weapons, munitions, and schools for the Communist officers. Shortly after the close of the Korean War, the Viet Minh defeated the French at Dien Bien Phu, and that phase was over.

It happened then that a hundred or so American advisers were sent to Vietnam to try to recapture the Western flag of colonialism. One of the advisers of the late 1950s

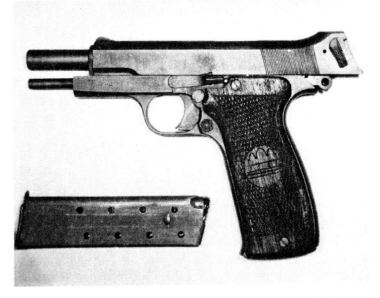

A Cambodian copy of the French M1950 pistol, liberated during action in Southeast Asia. Note the workmanship displayed in these three views of the weapon. (Courtesy of USAREUR Technical Intelligence Center.)

and early 1960s, then Capt. Lucien McCoy, reported: "Homemade and garage-sale weapons are killing people as well as the best issue our military has to offer to our side [South Vietnam]."

Frank Moyer, a retired Special Forces NCO and former ATF firearms expert, recalling his Korean War experiences, reported that improvised/modified weapons prevalent in that theater showed up several years later in

VC copy of a .45 ACP pistol, left side. (Courtesy of Maj. Richard Keogh.)

Kit Carson Scout with Frank C. Brown, Vietnam, 1971. The Vietnamese, a former Cong soldier, is carrying a cut-down, locally converted M2 carbine. (Courtesy of L. Caber.)

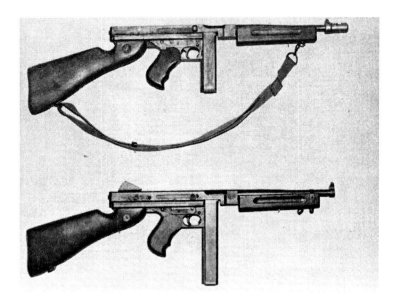

At top, a VC Thompson copy with original sling. The bottom item is the real thing. In the copy, note the imprecise cut on the cocking slot, the missing fin on the aluminum compensator, plus other assorted 1A1, M1928A1, and homemade parts. (Courtesy of Fred Rexer, Jr.)

Vietnam. Among the unusual items he saw was a twin set of M2 carbines mounted in a single stock and rigged to fire together.

"Sights weren't used," he stated. "Firing the weapon was mind-boggling, to say the least. It was somewhat less than accurate. But I suspect the psychological effect for both shooter and target would be stupendous.

Special Forces vet Fred Rexer, Jr. in Vietnam with the homemade VC Thompson he liberated. (Courtesy of Fred Rexer, Jr.)

"The guns didn't fire in unison, either," Moyer continued. "One would get ahead of the other, which made accuracy impossible and wasted a lot of ammunition. It was a half-baked idea, but it surely poured out the firepower. In retrospect, it wasn't much worse than some of the crackpot things we came up with!"

Former U.S. Marine R.S. Scott remembered capturing a Viet Cong officer and being curious about the man's rifle. "First, it seemed to be an issue M16," he recalled. "Then I looked more closely and saw that the rifle was fabricated from issue gun parts added to soft metal fittings, coloration, and such. But it shot fine.

"We put it up on the board of the club when I was able to smuggle it back from the field. Isn't that amazing, some little turd made that gun by hand to dust us Americans.

"Close up, you couldn't notice that the weapon was a copy. It also worked, but always on full auto. It was amazing how detailed that VC copy was," he added.

Fred Rexer, Jr. displays his VC-made Thompson SMG copy, captured in Vietnam. The weapon has no rear sight. (Courtesy of Fred Rexer, Jr.)

As a senior military advisor, Col. Frederick Roseman was allowed to bring an enemy gun home from Vietnam as a war trophy. Instead of the many flashy weapons, he chose a battered and cut-down old Dame shotgun.

Back in the States, he worked on the little 20-gauge, refinishing it into a fine and lovely sporting double barrel. What started out as a French plantation owner's sport gun was first converted for humid jungle warfare by the Viet Minh, then passed along to the Viet Cong for use against the Americans. Now Colonel Roseman has Americanized it as a fine sporting gun once more.

Hard work by nonprofessional designers brought our side interesting results, too. Pat Scrufari, an amateur engineer, modified a standard Smith & Wesson pistol into a double-barreled survival weapon. Scrufari's weapon is an over/under, with a .22 rimfire on top, bottomed by a .357 Magnum. The cartridges are nested in the same cylinder, with the .22s ringing the outer edge between the .357 holes. The pistol holds five shells of each caliber.

In 1976, Scrufari tried to interest the military in his

Pat Scrufari's over/under modification as tested by the U.S. Army as a survival pistol. The top of this heavily altered Smith & Wesson is .22, the lower a .357. (Courtesy of Pat Scrufari.)

The cylinder of Pat Scrufari's .22/.357 survival pistol with the smaller caliber nesting outside the .357 holes. (Courtesy of Pat Scrufari.)

unique survival weapon design, but the budget-bound brass said no.

Not all of the improvised/modified weaponry news came from Vietnam, though. In the earlier books of this series, John Minnery and I related the vital role of these homemade guns throughout the military history of the

Philippines. Despite the continued flow of regular military weapons from a variety of sources, including Vietnam, some antigovernment rebels still use home workshop firearms.

Lt. Randolph Johnson, a U.S. Navy intelligence officer, told me about local police in one of the small Philippine islands to the south confiscating two homemade submachine guns. Produced from plumbing pipe, sheet steel, and the battered parts from original Thompsons that must have been 50 to 70 years old, the guns were lethal. Lieutenant Johnson said the weapons "worked just fine . . . We took them to our range and they ran off a string of rounds as slick as the real thing."

Another ordnance expert, retired Marine CWO-4 Thomas Swearengen, has gathered additional information on some of the Philippine homemade firearms presented in the earlier volumes. Apparently, some of the Philippine weapons shown earlier may have been manufactured originally by Squires-Bingham. This firm, located in Manila, produced commercial revolvers, shotguns, etc., until the Marcos declaration of martial law, when all production ceased. Apparently, though, some folks within the plant began clandestine production runs of weapons on a limited basis to keep the anti-Marcos battle going. With the defeat of Marcos, of course, the plant was reopened and now continues to manufacture and export commercial firearms for the domestic and world markets.

In addition, Swearengen noted that the Moros who were fighting Marcos's troops were armed with about every firearm one might imagine. For example, he reported that "The Moros built shotshells from brass and steel tubing. The head was turned on a lathe and the shell body cut from the tubing, then pressed onto the head. This entire ordnance-gathering operation was in the planning stage for a number of years . . . since World War II."

The history of the Moros and armed insurrection goes back much further, though. They gave the Spanish a difficult time throughout their colonial rule of the islands.

Being fiercely independent and violently nationalistic, they gave both sides a fight during the Spanish-American War, often with homemade firearms.

Swearengen wrote: "This is where John Pershing got his initial fame and fortune with his bosses—way ahead of the Pancho Villa thing. Teddy Roosevelt liked the way Pershing handled the situation in the Philippines and promoted him way over many officers senior to Pershing."

The Moros were also a violently active boil in Japan's occupational backside during World War II. Many of the brave guerrilla bands active during the heavily publicized fighting were Moros with homemade or modified firearms. And why not? They had been fighting for their independence for nearly a hundred years by that time.

As a retired U.S. intelligence officer who knows that Philippine political/military situation well told me: "I'm sure the success of the Viet Cong made a big impression on the Moros. They realized our country [the United States] was backing another corrupt dictator and that those [Vietnamese] rebels actually won. No, the Moros are not Communists. Simply, they are people who want to be left alone. That's why they'll probably win, and I hope to hell we don't stay mixed up on the wrong side of this mess, too, like we did over there [in Vietnam]."

6 CHAPTER SIX

The CIA's Deer Gun

DREAMED OF AS THE VIETNAM generation's version of World War II's Liberator pistol, the CIA Deer Gun was like many improvised weapons, a crude, single-shot, 9mm pistol designed to bring better weapons to its users. R.D. Meadows, a former U.S. Army Special Forces officer, explained the concept.

"The idea was to air-drop or otherwise supply these glorified zip guns to our friendlies who carried the war to the enemy. They'd get close to a baddie, zap him with the Deer Gun, then strip him of everything usable, including his AK-47," Meadows explained.

The genesis of this weapon came during discussions among the military, CIA officials, and several gun designers in the late 1950s. The goal was a new version of World War II's FP45 Liberator, designed for the same mission. For the same insane reasons that other World War II materiel was destroyed in 1946–1947, Liberator pistols were torched, melted, and crushed into postwar scrap. Very few survived, and there was no inventory.

By the early 1960s, with covert operations already heating up in Southeast Asia, CIA procurers got together with the late Russell J. Moure, chief engineer for

American Machine & Foundry's special firearms division, to plan the Liberator's successor.

A legend in the military and paramilitary firearms business, Moure had created the classic AMF military suppressor, plus other ordnance designs for that company and The Company. He also designed the highly successful .308 drop-in conversion unit for the M1 rifle, as well as an 81mm semiauto mortar.

When the CIA's armament people met with Moure in 1962, the idea was to design a light, cheap Liberator-style pistol. The government's primary criterion for this new weapon was operational simplicity, followed by the ability to build it cheaply and quickly. The new pistol would supply indigenous guerrillas and irregular forces behind enemy lines.

One of Moure's engineering colleagues at AMF said: "Russ spent ten thousand words explaining to some CIA guys what was basically a crude, ugly, but damn decent four-dollar zip gun for our Third World allies to use to kill bad-guy soldiers. Then, to take that guy's weapon for his own use. That was the CIA program for this weapon."

Moure's design was a crude pistol with a cast aluminum receiver, a screw-out-to-load, two-inch barrel, and plastic parts, all of which was budgeted so that the weapons would cost the United States only $3.95 each. It was also totally sterile (i.e., there were no identifying markings). All component parts for the production guns were to be fabricated from nondomestic sources outside the United States for further "operational denial," which was how the CIA described its attempts to hide the weapon's origin.

Each weapon measured 5 inches in length and was 4 1/8 inches high, 1 1/2 inches thick, and weighed 12 ounces. It had a blued barrel and a bright aluminum handle/receiver unit. The grip was hollow to hold spare ammunition and an ejector rod to punch out the empty casing from the screw-off barrel. Obviously, reloading was not quick-'n-easy for this one-shot assassination weapon.

The Deer Gun had no fixed sights and was fired by

The legendary ordnance designer Russell J. Moure, shown here with a ballistic knife, developed the Deer Gun on a CIA contract. (Photo courtesy of Jack Krcma.)

means of a cocking knob and trigger unit. The only safety was a plastic ring that slipped over the cocking knob, acting as a collar to prevent it from falling on the cartridge primer if the weapon were discharged accidentally. This collar also doubled as a front sight, to be slipped off the cocking knob and attached to the barrel before firing. After examining and testing Moure's prototype, the Agency ordered 1,000 pistols, issuing AMF a developmental contract for $300,000, quite a bit higher than the original price of less than $4 per weapon.

"These were developmental weapons, which meant our research and prototype costs had to be recovered," explained the former AMF officer, who was on the Deer Gun project with Moure. "These costs were high, but if we'd gone into mass production the projected figure of $3.95 could have been met."

Each pistol came packed in a plain white polystyrene box along with three rounds of 9mm ammunition, sterile of any markings (i.e., head stamps). The box also contained a four-color, cartoon-style, wordless instruction sheet that visually and in detail instructed the user how to operate the weapon and upon whom.

The cartoon instruction sheet depicted a generic guerrilla using a Deer Gun to shoot an enemy soldier, who was wearing a Soviet hammer-and-sickle arm band. Ironically, the only identifying mark of any kind found on the weapon, its container, or the instructions was that arm band.

Of the weapons delivered to the CIA, approximately 150 were sent to Southeast Asia for field-testing, according to a consensus of sources. Although there is no official record that any of these pistols were used for anything beyond controlled, noncombat testing, one U.S. military officer disagrees, saying he had accompanied a patrol of both U.S. and Vietnamese Special Forces in which the Deer Guns were carried for "active evaluation," as he put it.

The rest of his story is a tad grisly, but I have no reason to doubt the man's word. I've known him and his honesty since 1960. Here's what he told me:

"We had run a successful ambush and were returning for extraction with four prisoners, three of whom were wounded. The unwounded man was hostile and resisted restraint. Because of the surroundings and because the potential for hostile reaction to us being there was very real, our senior man decided to terminate the recalcitrant prisoner.

"That's when I saw the Deer Gun 'field-tested.' It was used to terminate the prisoner. One shot was fired from a range of 3 feet into the back of the base of the man's head.

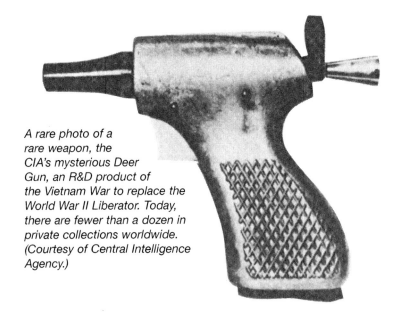

A rare photo of a rare weapon, the CIA's mysterious Deer Gun, an R&D product of the Vietnam War to replace the World War II Liberator. Today, there are fewer than a dozen in private collections worldwide. (Courtesy of Central Intelligence Agency.)

THE CIA'S DEER GUN • 85

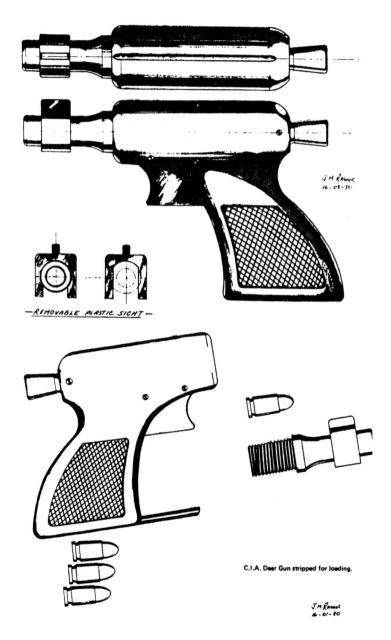

C.I.A. Deer Gun stripped for loading.

Schematics show the CIA Deer Gun as issued (top) and stripped for loading (bottom). (Courtesy of J.R. Ramos.)

He lurched forward and fell. He was pronounced dead immediately. We then effected our extraction."

By 1964, the Deer Gun was listed in the CIA's special weapons inventory and carried a regular stock number, meaning it was cleared for field issue. That stock number was 139-H00-9108.

Several researchers and experienced military operations people have speculated on the origin of the Deer Gun name. Sgt. Gary Paul Johnston suggests that it is an Agency code name with sardonic reference to a survival weapon. Suppressor designer Don Walsh, a longtime friend of Russ Moure, thinks the weapon was named after a World War II OSS operation, the mysterious Deer Mission business in Burma. The genesis of their name appears to be as enigmatic as the ultimate fate of the weapons themselves.

Of those original 1,000 weapons, only 20 to 25 remain in circulation today, according to collector/historian Keith Melton. The majority were destroyed by the government through orders reportedly issued directly from the White House, Melton and others feel. The rest seem to have just plain disappeared, meaning, probably, that they were liberated by individuals.

After AMF, Russ Moure went to work for Firearms International as chief engineer. Later, he joined Interarms as vice president of engineering. Moure died during the winter of 1986–1987 when, after stopping to help a stranded motorist near Washington, he was struck and killed by another vehicle.

7 CHAPTER SEVEN

The Fringe's Death Benefits

AN ANONYMOUS TELEPHONE CALL alerted a South Florida police organization about an abandoned shotgun that the caller "had seen" sticking in the Dumpster behind a store in a shopping mall. Answering the call, an officer opened the action of the shotgun to check for shells. When that action was opened, a wire pulled the free contact onto the fixed contact, completing the circuit and detonating a bomb, killing the officer instantly and injuring his partner.

After combing the rubble, forensic investigators concluded that 6 ounces of low-tech explosives were hooked to an electrical blasting cap, three 1.5-volt batteries, a plastic hair curler, two foil contacts, and connecting wire. The device was fed into the barrel of the shotgun.

The dead officer's captain said, "What in the hell kind of animal leaves a hideous death trap like that?"

The captain already knew the answer: a very nasty South American drug warrior who wanted to terrorize the local police, reminding them that he was operating his drug war in their territory and that nobody was safe.

Another form of modified shotgun was reported in Florida, Texas, and in New York in 1990. It was a cut-down 12-gauge shotgun modified to fit into the interior of a car

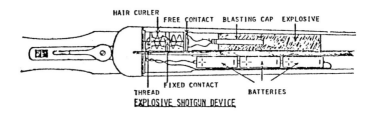

A lethal shotgun booby trap has haunted police throughout the Eastern U.S. First discovered in Florida, the device consists of 6 ounces of commercial explosive, an electric blasting cap, three 1.5-volt batteries, a plastic hair curler, two foil contacts, and connecting wires, all contained in the two barrels of a shotgun. When the weapon is broken open, the circuit is completed and the device explodes. (Courtesy of Pennsylvania State Police.)

door, where it was mounted with the barrels pointing toward the rear of the vehicle. The weapon is fired by a trip handle or wire handled by the driver, who opens the door slightly to aim the weapon rearward toward the target approaching the vehicle. The main target for this sort of ambush, obviously, would be police officers, customs officials, or DEA people.

"The armed fringe of our increasingly violent society is out there getting meaner and more deadly diabolical all the time, especially as lawful control either closes in or is perceived as doing so," terrorist expert Brian Jenkinson told me, adding that he felt that the political, economic, and social turmoil of the twentieth century would escalate this carnage from the violently lunatic fringe of both the Left and the Right.

In the United States today, serious law enforcement people recognize that the major threat of terrorism comes from the Right fringe rather than that of the Left. This despite 10 years of political bleating from the White House, which has turned much of America into a police state where ordinary citizens, fearfully and stupidly, have given their Constitutional rights away for vague political promises of better protection from street crime, drugs, leftist terrorists, Arab terrorists, porn terrorists, etc.

In truth, according to most intelligent law enforcement professionals, the extremist groups from the Right are almost always better led, equipped, and financed. Many of their disciples have military or police experience. And they almost always outgun today's Wobblies of the Left, not to mention law enforcement folks.

Sometimes my work jumps off the research page and grabs me by reality. I have seen war, politics, terrorism, more war, crime, and assassination firsthand, on the scene, in and out of the United States. Several years after finishing my book on terrorism, for which I had interviewed several terrorists in prison and also some on the run, I got a letter from a "retired terrorist . . . an anarchist who's given up." He said two woundings and a stretch of time in prison had convinced him to give up the fight. He wanted to tell me some of his war stories.

At best I am conservative and cautious about this sort of contact; at worst I am cynical. Yet the adventurous side of my brain won, and I agreed to meet the man. Our first meeting was an exchange of bona fides and credentials, such as they are in this business. In some respects this reminded me of two dogs sniffing about each other.

Our second meeting happened after we were both satisfied with each other's experiences, honesty, and sincerity of purpose. Today, knowing what has happened since that I didn't know then, I am certain this man was telling me 100-percent truth, which makes what's about to follow all the more frightening. After our two sessions, I never saw him again, and to this day I have no idea where or what he is, or even if he's still alive.

What follows are selected versions of our conversations, edited mostly for their relevance to the topic of this book, improvised/modified firearms and their use.

"I became involved in rightist paramilitary activities at an early age [16]. Later, when things looked more serious, I used legitimate and other contacts to acquire whatever we wanted for my group.

"It is like a developing band of assassins who have prepared their equipment well, trained, and only await the right leader to give the word or wait for the United States to get up to its ears in so much internal and external turmoil that it cannot function effectively . . . Then, we of the Right will emerge in small bands to take power.

"Griffith Park in central Los Angeles was our training area for awhile. It closes at 10:00 P.M., and we would go in shortly after closing in groups of a dozen or more. There'd be two five-man teams and instructors, and we'd pursue a paper training scheme, such as taking dummy cases of dynamite up to the police communications building and tower. We would go through the motions of blowing up the place. Usually we went in fully armed and were actually ready for anything.

"During this time there were rumors of black groups doing exactly the same thing we were doing, so we literally went hunting for them in Griffith Park . . . Never found any.

"We had some impressive weapons: M1 carbines, a couple of Ruger .44 Magnum carbines, some AR-15s, and a couple of DEWAT (deactivated war trophy) Sten guns that a machinist activated for us for 10 dollars each. We had

This crude copy of a Mauser pistol, confiscated by police officers in Los Angeles, was probably made in Red China, according to firearms expert Art Wesley. (Courtesy of Los Angeles Police Department.)

homemade silencers for our pistols, usually made from lawn mower mufflers, which we drilled out . . . It was easy to get guns on the street or steal them . . . or make them or modify what we could get. We were pretty good gunsmiths.

"During this period, there were a dozen or so people working with me and quite a few loyal machine shops, so we could contract individual parts without anyone else knowing what the finished product would be. One old machinist did all the fine work for patriotic reasons, and on him I depended for shop drawings and specifications. We would rent time in several loyal local shops for work we could not handle ourselves. We built and converted machine guns and silencers this way.

"In the course of developing our silencers, there was no real scientific research or engineering. We simply set out to limit the noise factor to acceptable levels for use in urban situations.

"Most of the silencers and other pieces of military armament we produced went to such groups as the Minutemen, the California Christian Rangers, the Christian Nationalist Army, and for us, too, et cetera."

*****

"We were a hard core of radicals who were responsible for at least 95 percent of all the heavy racial and political violence in California. We used all of the groups and other political organizations such as the John Birch Society. I was a captain of such a group, and my team numbered 50 people. The police and federal authorities were well aware of our activities and know who is involved in what. But there is no way of proving anything on anyone."

*****

"Gun laws are a silly-assed crock of shit. They simply did not apply to us. What we couldn't buy or steal, we made. We took lots of regular guns and easily made them into machine guns. We made explosives and grenades and silencers . . . We took a couple of Browning pistols and converted them into full automatics . . . total killing machines for close-in assassination and murder."

He concluded by telling me that, as far as he knew, many of the people and most of the weapons are still out there . . . "waiting for the right time and moment . . . Maybe that time is coming soon. I'm out of the game, but plenty of the guys and gals are still in place, and so are their homemade machine guns . . . and, baby, they work," he concluded.

*****

In addition to my independent verification of this man and his story, headlines support what he told me. For example, a *Los Angeles Times* story reported a seizure of silencers at the home of a man who was booked on illegal weapon possession charges. Police recovered 23 silencers in various stages of assembly, plus five more that he had attempted to sell to an undercover agent in an L.A. parking lot.

Another story related the seizure of 15 homicide machines in Glendale; six more turned up two months later. How many more are out there?

The U.S. military had an unwilling hand in arming the fringe, too. Late in the summer of 1975, Les Aspin, a Wisconsin congressman, forced public disclosure of a document which had been classified "Top Secret" for "national security reasons." The official Defense Department document reported that "unknown persons" had stolen a total of 6,900 military firearms and 1.2 *million* rounds of ammunition over a three-year period alone. All the thefts took place in the continental United States. Such a quantity of arms and ammunition is enough to equip 10 battalions of troops.

Asked if there was a similar situation in overseas commands, Pentagon officials told Aspin in public hearings, "We have no way of knowing."

One former member of the Weather Underground,

interviewed late in 1980, related: "One of our brothers was a corporal in Vietnam during 1970. He managed to get us enough parts to put together six M16 auto rifles. I was told that, earlier, a lot of guys were mailing back grenades, Tommy guns, and all that good stuff. I know we had some heavy stuff."

He said that they used the existing weapons as models for the construction of other automatic weapons, produced in home shops, adding: "We had experienced, military-trained armorers and some engineering kids. They'd take one of these stolen army automatics and use it to help us turn out homemade guns in some friendly shops we ran back then."

Asked what had happened to these weapons and their owners from back in the 1970s and where they ended up, he smiled, shrugged, and said, "Beats me, that's The Man's problem."

Fast-forward to 1991 and to troops coming home from Desert Storm. According to four officers whom I interviewed separately back in the States under their own strong security-of-identification restrictions, hundreds of weapons disappeared "God knows where." Despite almost draconian gun control measures, hundreds of M16s, FALs, and AK-47s have ended up in the United States illegally.

"We're kidding ourselves if we think the majority ended up in some grunt's basement as a harmless war trophy. A lot of very bad people now have illegal automatic weapons," one field grade officer who is a Criminal Investigation Corps (CIC) investigator told me. "Every time we have a war, the lunatics get a bunch more weapons, and we've trained a bunch more lunatics to create, fix, and fire homemade machine guns as well."

These deadly problems are not limited to the United States. For example, in South and Central America, radicals on the Left and Right wage continuous underground war against the government and each other, depending upon who is in power. An AP wire service story reported: "Argentinean Federal Police uncovered a $900,000 underground arms factory hidden on a farm near Buenos Aires in April 1991. The police claim that drug cartel-financed guerrillas of four nations were working to perfect a cheap homemade light machine gun design."

The guerrillas had dug out a large basement room beneath a barn in Moron, a village 30 miles from the capital city. Among the 25 terrorists arrested were natives of Chile, Bolivia, and Uruguay. The rest were members of the outlawed People's Revolutionary Army of Argentina. Among the 25 arrested were 15 women, who were working to produce small arms and automatic weapons.

According to noted firearms expert Thomas Swearengen, USMC (Ret.), the Beretta selective-fire 951R is a prized possession among Italian terrorists. He also reports that some of these outlaws have converted standard 951 pistols to the fully automatic function. Both the standard 951R and the converted models were reportedly used in most kneecapping attacks and assassinations in that country for the past 20 years.

When I was in Guatemala in 1986, I saw a local army intelligence report about a cache of stolen guerrilla weapons recovered near the village of Quiche. Although most were stolen from the Guatemalan army, one of the M16s had been stolen from an American National Guard armory in Texas, while another had been left behind in Vietnam. There were also two large fabricated shotguns, which fired homemade 4-gauge shells.

An American military adviser who showed me the report told me that he wanted the local army to arm some of the *campesinos* so they'd have protection against these roving bandits as well as the guerrillas. The Guatemalan government refused.

He laughed at the memory, then said to me: "What's that tell you about the government here, when they're afraid to give guns to their own people?"

He added that while he had no time for the philosophy of the Sandinista government in Nicaragua, they had at least armed their citizens because everyone belonged to the militia, and the government trusted its citizens.

"The Guats won't arm their *campesinos* . . . It's a dictatorship. So the locals armed themselves," he said. "One man makes 10 barrels for these crude, single-shot 12-gauge guns from plumbing pipe. Another old boy shapes bolts, while another makes stocks from the native Quiche pine. One of the smarter ones has made a trigger group from springs and pins and odd bits of metal. Then they gather and assemble guns to defend their settlement and to use for hunting."

I asked where they got ammunition and was told: "Sometimes a case or so of our 12-gauge field loads might fall off the back of my jeep . . . someone's gotta help these poor bastards if their own government won't."

There was one humorous bright spot in all of this. That would be the attack of the deadly nail guns in 1982, when the New York news media, never a boring or objective lot at best, went ballistic following a freak accident in a local restaurant, where a female construction worker was injured when struck by a nail driven by a power nailer. The *New York Post*, that city's version of the infamous tabloid, *The National Dumper*, editorially shrilled about the "easy availability of deadly cartridge-fired nail guns . . . can be obtained with nightmarish ease by anyone who wants one."

With language bordering on hysteria, as if the entire population of Crips gang members, all armed with the deadly nail assault guns, were looting and raping their way through the five boroughs, the *Post* described how one of its reporters went out on a serious undercover mission . . . to rent a nail gun. And, in this episode of award-posturing investigative journalism, the reporter managed to rent one on her first try. The *Post* continued in prose I could not have invented:

> This deadly assault weapon would shatter a block of concrete and drive a nail through 8 inches of weathered pine . . . The ease with which they can be obtained, at hardware stores and tool rental firms, makes a mock-

ery of state and city gun control regulations . . . they are a mugger's dream come true . . . When not firing nails they can easily be converted into a zip gun with materials found on the average handyman's workbench, and owning one is perfectly legal.

According to several New York City police information officers and a safety officer with whom I spoke, there have been no reported incidents involving either nail guns or converted nail guns before or since the 1982 editorial campaign by the *New York Post*. But, apparently, one Hollywood scriptwriter with a sense of humor may have been influenced by all of this, if you recall the scene from one of the *Lethal Weapon* films in which costar Danny Glover wasted a bad-guy hit man with a nail gun. All in all, the entire issue seems to be very tacky.

8   CHAPTER EIGHT

Barring All Guns

YEARS AGO, AMBROSE BIERCE WROTE that prisons were places of punishments and rewards. He added that "the poet assures us that 'Stone walls do not a prison make,' but a combination of the stone walls, the political parasite and the moral instructor is no garden of sweets." A prison is an institution of plotting and planning, as well as a graduate school of crime. It is a highly regulated and controlled society, and it is also a heavily armed society, on both sides of the barred cell door.

Prisons are a paradox. Weapons flourish where there are to be none. Inmate ingenuity has given rise to weapons made from the most unusual and mundane items. Late in 1977, for example, Mexican national police made a thorough sweep/search of the infamous Jalisco State Penitentiary after a riot over inhumane conditions there. Nearly 1,500 weapons were confiscated, including 30 pistols and two submachine guns, which had been built in the prison entirely from fabricated parts. Stories like that make a total mockery of the contention that antigun laws keep firearms out of the hands of criminals.

A 1982 article in *Gun Week* quoted an inmate in a Michigan prison who actually had a price list for buying

Leo Jenkins, former warden of the Indiana State Prison, examines some of the contraband weapons found at that institution during a search. He is holding a prisoner-produced pipe pistol. (Courtesy of William Swendenberg.)

A close-up of some of the weapons seized at the Indiana State Prison, including a number of pipe and zip guns. (Courtesy of William Swendenberg.)

weaponry produced inside the walls. He said that a real gun is expensive, "$500 including ammunition for a piece that on the outside would go for $150." He said that homemade guns were only $350, while a shank (knife) was "easy, $5 for a 6-inch blade and up to $10 for a 10-incher."

Guns and the prison escape scenario run the gamut from John Dillinger's soap-carved gun to zip guns to high-tech electronics. In 1985, an inmate tried to power his way out of a Wisconsin facility with two zip guns made in the prison, which was described in an interview with an inmate as "a very, very to the maximum . . . maxed-out and watched place." The prisoner told authorities only that he had bought the homemade .22-caliber guns, which were made by another prisoner inside the institution, while the ammunition had been smuggled in to him.

A very unusual prison escape made headlines late in 1975, when inmates from the Indiana State Prison at Michigan City broke out using a homemade, yet sophisticated electrical device that opened all the power-controlled gates for them. This happened not too many months after a detailed weapons search turned up quite a haul.

For example, the authorities' efforts turned up six pistols made from plumbing pipe parts, several .38 and .22 cartridges, and 20 homemade knives. Commenting on the find, *Gun Week* expressed surprise that commercially produced weapons were not found. They reported:

> Normally, in the experience of most prisons in this country, such searches turn up at least one, if not several factory-made firearms . . . none of the firearms located in this search were factory-made . . . But, the homemade models were .22 caliber.

Their report noted the irony of homemade firearms showing up in a prison society with its stringent security precautions, noting especially that this particular prison had been under close security for some time due to earlier escapes. The weapons were apparently made during this security period.

This single-shot tip-down pistol was confiscated from a convict in transit between prisons. It showed real craftsmanship, including a floating firing pin and an unrifled barrel finished so well that it shot .38-caliber cartridges accurately. (Courtesy of Leslie L. Smith.)

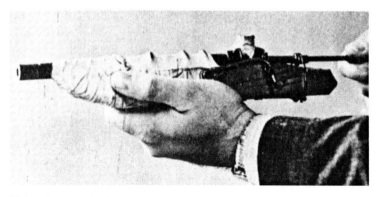

This prison-made, break-top-type .410-gauge shotgun was fired by pulling back on the spring and releasing. (Courtesy of Clair F. Rees.)

As most any prison-knowledgeable person will tell you, the majority of illegal weapons held by prisoners are strictly for defensive purposes. The majority are small, homemade knives, but homemade guns are not uncommon. Every month, in prisons all over the world, zip guns turn up

A San Quentin convict fabricated this pistol while an inmate, a point California State Senator Bill Richardson made to demonstrate the error of gun control as a workable manner of fighting crime. (Courtesy of Sen. H.L. Richardson, California.)

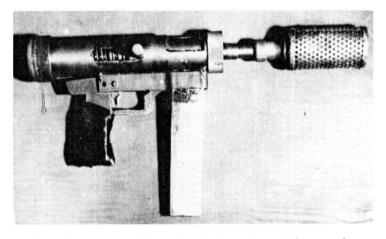

A silenced weapon used by an inmate in an attempted escape from San Quentin. (Courtesy of Capt. H.A. Tobash, San Quentin State Prison.)

when lockdowns and tosses happen. Most are crude devices made from pipe, wood, springs, and tape. Most fire a single .22 round.

Some years ago I played for a softball team that made regular visits to a state penitentiary to play an inmate team that participated in our local recreational league (theirs were all home games, I might add). The inmate team was always good because it didn't have the job mobility attrition problem that plagued our outside teams. Over the course of several seasons, I got to know some of the guys well enough to share some talk. I often asked about weapons, and because I had the trust of a few veterans of the system, they answered honestly.

"There is a lot of nasty gang shit in here, mostly racial and where you're from [ethnic]," one lifer told me. "If you don't belong to a power group [gang], you'd better be a shadow or a physical power . . . Or like me, part of the permanent furniture so I'm let alone.

"Some of the younger, weaker ones buy zip guns made inside. That's true for the babies and the mamas (young male prisoners soon to be bait for nasty homosexual rape). It's a nasty place in here. Some gotta have a weapon just to survive, 'cause the guards don't care."

He described the guns he'd seen as "the usual zip guns like we had as kids on the street. You know, a piece of pipe to fit the shell, a wood handle, and a spring or rubber band to drive a nail into the bullet to fire it. You gotta put that kinda gun up to the man's face to make it do any good. But, I tell you, men in here have them guns, they sure do."

As my friend related, there is ample empirical evidence to support the claim that most prison firearms are held for defensive purposes. In the fall of 1975, criminologist Gordon Firman conducted research among a sampling of Ohio prison convicts, whose consensus was that use of a firearm to prevent attack on one's person or property was justified. The majority also "had easy access" to homemade guns in prison.

Firman also reported that prisoners scoffed at the idea that confiscation of guns from civilians would do anything to decrease the outside/civilian crime rate. Most convicts felt that assault crimes and daylight burglaries would increase dramatically if citizens were disarmed.

During prison riots, shop-generated firearms are also seen. Videotapes of prisoner riots in Iowa, Pennsylvania,

Michigan, and New York in the 1980s and 1990s show riot leaders armed with small, deadly zip guns made inside the walls. Those guns recovered after order is restored are usually of the pipe-and-spring, .22-caliber variety, although one homemade weapon taken in an Iowa prison riot was a smoothly produced, single-shot 9mm pistol.

A prison spokesman noted: "It was strong, well-designed, and tightly built. The action was smooth and clean. The man who put this together is very talented. It's a shame his talent is not more positively channeled."

I think that view is tempered by which side of the cell door you walk.

In the late 1980s, an enterprising newspaper reporter investigated a thriving firearms business in Graterford State Prison in Pennsylvania. He wrote: "Prisoners were buying small, crude pistols built from scrap material right in the prison itself by other prisoners. . . no laws, no waiting period, no paperwork . . . It is gun control at its obvious worst."

The reporter noted that the majority of prisoners wanted weapons for self-protection and that some did keep homemade guns hidden for just that purpose. Prisoners told him that guns cost between $300 and $500 each, depending upon quality of workmanship.

In northern California, the *San Francisco Examiner* noted that "the manufacture of firearms in San Quentin maximum security prison has kept some inmates quite busy." The story quoted a San Quentin official as saying that they had confiscated "dozens of homemade guns, gunpowder, and other weapons during an internal investigation of illegal workshop activity."

The prisoners reportedly made gunpowder by taking sulfur from match heads, charcoal from burned wood, and nitrates from dried urine. This propellant is used in their hand-held guns made by coating magazine pages with soap and water, rolling them into tubes, binding the tubes with string, plugging one end, and drilling a hole for a fuse.

The fuse is made from toilet paper soaked in dissolved match heads. Projectiles can be anything from ballpoint pen points to small nails to pebbles.

San Quentin guards reportedly confiscated an average of 20 to 25 weapons a month, according to the official quoted by the *Examiner*.

This prisoner arms-manufacturing industry obviously reflects an increasing awareness among criminals about how to circumvent restrictions on access to arms. In our prisons, these activities are especially ironic, since guards patrolling inside usually are unarmed.

Putting the concept of easy access to firearms in a prison environment into perspective, retired federal probation and parole officer Wallace Richard Croup said: "You'd think a prison would be the safest place on earth and a model for gun control. If you thought that, you'd be 100-percent wrong, and, if you were inside and thought that, you would be dead."

That challenging truth should give you far more than philosophical awareness of the prospect of an in-prison shotgun marriage, or divorce, with Bubba Ray.

CHAPTER NINE

Meanwhile, in the Basement

WITH THE EXCEPTION OF SERIOUS fabricators of improvised/modified firearms whose missions, safety, and even lives depend on these weapons, the majority of designers and builders are that good old American buddy, the home workshop tinkerer. While their manufacture is not quite a cottage industry, these firearms hold legitimate station in the American tradition of home gunsmithery.

Some of these gunsmiths are fiercely independent Americans, such as one who calls himself Steve Higgins, which obviously is not his real name, as that was, at the time, the real name of the director of the ATF. The ersatz Mr. Higgins says: "What I make in my own home is my own business so long as I don't harm anyone else or threaten to do so. The feds and their gun police can go [commit sodomy on] themselves . . ."

Others are like Allen Freed, a hunter and competitive shooter, who says: "I just enjoy tinkering with old guns and trying to build small firing devices into nontraditional gun formats. It's no big deal, and they never leave my workshop . . . No, I have no license to build anything, and I usually destroy what I build. It's just fun, I mean no harm, and I don't want to spend the hassle or the money to get a federal license."

Some of these enterprising engineers do go limited commercial, though. In almost any issue of *Gun List* or *Shotgun News*, you see advertisements for various forms of improvised/modified firearms and/or plans for such weapons.

In the early 1980s, there was a company known as Covert Arms Manufacturing, located in El Paso, Texas, which sold commercially produced and properly registered penguns in a variety of sizes and calibers. Their products are covered in more detail in Chapter 11.

Today, however, the master designer for the home tinkerer is a fascinating professional who calls himself Richard Sardaukar, which is also the name of a character from Frank Herbert's novel *Dune*. He is the impresario of Sardaukar Press, located in Germantown, Tennessee, and publishes pamphlets and plans for building a variety of improvised/modified weaponry, including firearms.

A most interesting and charming gentleman, Sardaukar produces and markets detailed plans, schematics, and photographs of nine very exotic and interesting forms of improvised/modified firearms and other weaponry. He also

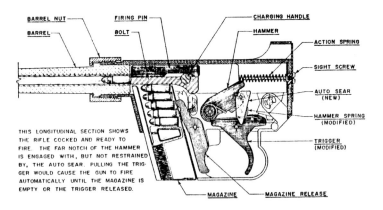

The Rocker. The longitudinal section in this illustration shows the rifle cocked and ready to fire. The far notch of the hammer is engaged with, but not restrained by, the auto sear. Pulling the trigger would cause the gun to fire automatically until the magazine is empty or the trigger released. (Courtesy of Richard Sardaukar.)

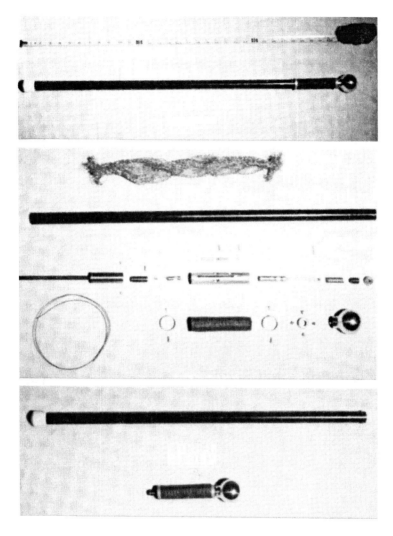

Top and right: Unlike the traditional caneguns of history, Richard Sardaukar's professionally engineered design goes one step further— it's also very efficiently silenced with a simple but highly effective sound suppressor built in. (Courtesy of Richard Sardaukar.)

CANEGUN

SILENCED .22 L.R. CAL. DEFENSE WEAPON

DISCLAIMER: THE DRAWINGS HEREIN ARE SOLELY INTENDED FOR THE EDIFICATION AND ELUCIDATION OF THE READER. SARDAUKAR PRESS DISAVOWS ANY RESPONSIBILITY FOR APPLICATION OF THESE DRAWINGS TO ANY PURPOSE OTHER THAN THE USE STATED ABOVE.

WARNING: IT IS A VIOLATION OF FEDERAL LAW SUBJECT TO FINE AND / OR IMPRISONMENT TO MANUFACTURE THIS OR ANY FIREARM OR SILENCER WITHOUT A LICENSE.

Sardaukar's version of the canegun. (Courtesy of Richard Sardaukar.)

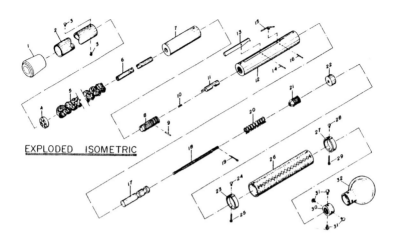

Half-size exploded isometric drawing of the Sardaukar canegun. (Courtesy of Richard Sardaukar.)

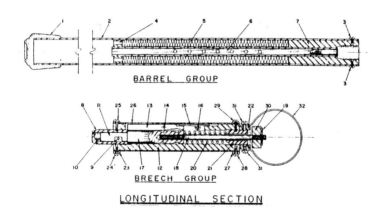

Full-size drawing of the Sardaukar canegun's barrel group, top, and breech group, bottom. (Courtesy of Richard Sardaukar.)

markets copies of ATF Form 1, the only way to fabricate your own weapons legally.

BOLTGUN

.22 L.R. CAL. DEFENSE WEAPON

DISCLAIMER: THE DRAWINGS HEREIN ARE SOLELY INTENDED FOR THE EDIFICATION AND ELUCIDATION OF THE READER. SARDAUKAR PRESS DISAVOWS ANY RESPONSIBILITY FOR APPLICATION OF THESE DRAWINGS TO ANY PURPOSE OTHER THAN THE USE STATED ABOVE.

WARNING: IT IS A VIOLATION OF FEDERAL LAW SUBJECT TO FINE AND/OR IMPRISONMENT TO MANUFACTURE THIS OR ANY FIREARM WITHOUT A LICENSE.

The Sardaukar boltgun. (Courtesy of Richard Sardaukar.)

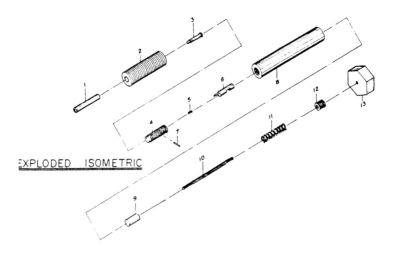

Half-size exploded isometric drawing of Richard Sardaukar's boltgun design. (Courtesy of Richard Sardaukar.)

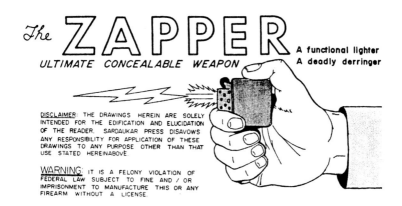

Sardaukar's Zapper: the ultimate concealable weapon. (Courtesy of Richard Sardaukar.)

The interesting improvised/modified firearms listed in the Sardaukar Press catalog include the following:

1. *The Rocker.* This clever drop-in auto sear allows you to convert a Charter Arms AR-7 into a fully automatic rifle with no alteration to the frame, bolt, hammer, safety, or firing pin. The new sear is described by Sardaukar as "a design refined over the course of the past 10 years to produce a conversion of unbelievable simplicity . . . basically a piece of bent brass and a drill bit."

Five detailed sheets of shop drawings give clear instructions on how to create a .22-caliber submachine gun with an AR-7 and this sear. The instructions cover everything in detail, from the theory of operation to the method of construction and operation to the manufacturer's original replacement parts. The design allows the weapon to be returned to its original semiautomatic configuration in seconds without any trace of the prior conversion.

2. *The Canegun.* This improvised firearm is a single-shot .22 Long Rifle walking stick weapon with a quiet difference: it features a built-in sound suppressor. According to Sardaukar, "The inherent design, length, and diameter of the cane itself houses a high-efficiency suppressor."

This home workshop canegun also features an inertia

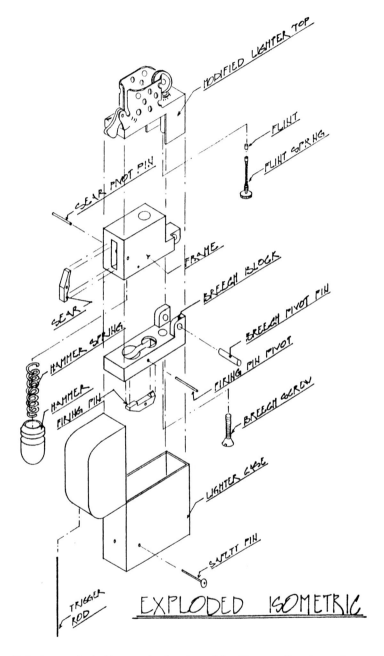

Full-size exploded isometric drawing of the Zapper. (Courtesy of Richard Sardaukar.)

MEANWHILE, IN THE BASEMENT • 115

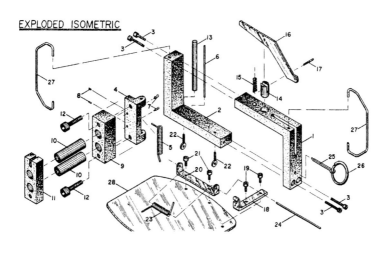

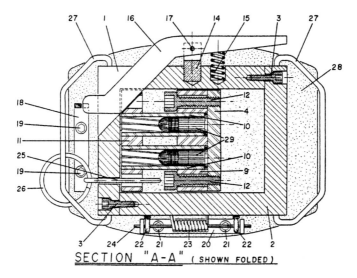

Exploded isometric drawings of the Sardaukar Buckler. (Courtesy of Richard Sardaukar.)

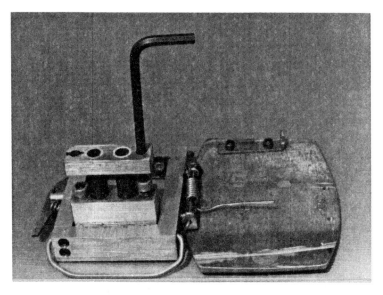

The internals of the belt-buckle gun are simple, yet professionally designed. The professional engineer who designed this device cautions that it is not a toy or a cheap replica. "It is a robust, top-quality firearm with great concealment—a major plus for self-protection." (Courtesy of Richard Sardaukar.)

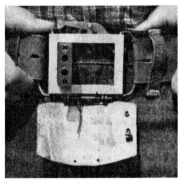

The Sardaukar Buckler is a double-barreled dealer of deadly belly-ache. It goes from being an ordinary belt buckle to a lethal two-shot .22 firing device that can be made easily in the home workshop. (Courtesy of Richard Sardaukar.)

MEANWHILE, IN THE BASEMENT • 119

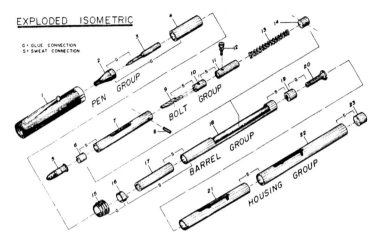

Full-size exploded isometric drawing of the various parts of the Sardaukar pengun. (Courtesy of Richard Sardaukar.)

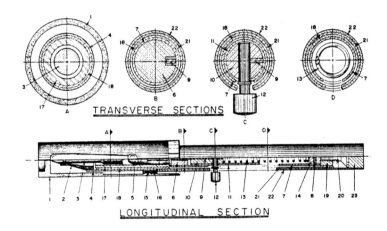

Exploded drawings of the transverse and longitudinal sections of Sardaukar's pengun design. (Courtesy of Richard Sardaukar.)

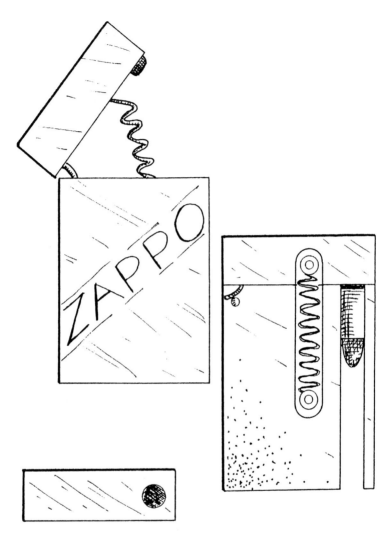

Polished steel monogrammed lighter pistol in .22 Short. (Illustration by Chris Kuhn.)

firing pin, a totally concealed trigger, and hammer rebound safety to prevent accidental discharge if the stick is dropped or used as a club. Sardaukar, inventor of the weapon, adds: "With the exception of the rifled barrel liner, available from Brownell's, all other parts are simple hardware store items. My primary objectives in this design, as in all that I do, are simplicity of mechanism and ease of construction."

The Sardaukar design uses minimal machining and comes with complete, illustrated, step-by-step instructions, including shop drawings of exploded isometric and orthographic views. Sardaukar adds, "This canegun, while simple to construct, is not a novelty or a toy. It's the real McCoy: a deceptive, quiet, and deadly firearm for self-defense use."

3. *Boltgun*. This improvised firearm looks and feels just like an ordinary 7/8" x 2" man-made bolt. Unlike an ordinary bolt, though, this one can really screw up your opponent; it chambers and fires a .22 Long Rifle round.

"The device is fired by holding the bolt shaft and pulling back on the bolt head, much like propelling the ball on a pinball machine. You just aim and let loose of the head to fire the device," Sardaukar explains.

An inertia-type firing pin prevents accidental discharge, he adds. The construction steps are contained in six sheets of illustrated and detailed shop drawings.

4. *Zapper*. Sardaukar advertises this one as "the ultimate concealable weapon," because it is a fully functional lighter as well as a fully functional and deadly single-shot derringer pistol. Sardaukar says, "Except for possible legal technicalities, you can carry the Zapper and be armed and undetected. It looks, feels, smells, and lights just as a lighter should." Of course, there is that one deadly difference—it fires that single .22 round.

As always, each kit comes with totally illustrated, detailed plans and drawings. Instructions note that the only power tools necessary are a drill press and a Dremel hand grinder.

According to early advertising, the original zapper began life as the Zippo Zapper but when Sardaukar got a nasty letter from the trademark owner of that first name, he quickly dropped it and has never used that first Z word again.

He says of his invention, "It's a unique weapon/lighter which can be carried with impunity to help you out in social predicaments, both light and serious, depending upon which function you require."

5. *Buckler*. The original of this device, the four-barreled German World War II belt-buckle gun, was described and pictured in the earlier volumes on improvised/modified firearms by John Minnery and myself. Sardaukar has made his improved model much smaller, lighter, and less obvious. The mechanism for his double-barreled .22-caliber firing device is a scant half-inch thin and is totally hidden behind a common flea-market-variety flat belt buckle.

The weapon is activated by depressing a single lever, which causes the buckle to flip down and the two barrels to snap out and fire simultaneously, perpendicular to your stomach. This is going to be very alarming news to whoever is standing in front of you, giving him or her an instant bellyache.

His plans for the Buckler include 11 sheets of illustrations and shop drawings, orthographic projections, plus parts and materials lists.

6. *Penguin*. The product of nearly half a year's research and development, this unit is totally unlike any other on the market. It truly is a real pen/gun, in that it will both write and shoot. It can be carried in your shirt pocket without arousing the slightest suspicion, especially when you pull it out and begin to write.

Unlike many so-called penguns, this one is not a hunk of cheap water pipe that looks crude and is quite dangerous to firer as well as firee. This is a professionally designed firearm which also looks, feels, and writes just like the fine pen that it is. This single-shot, .22-caliber firearm is built from common parts, most of which are fine brass tubes found in any good hobby shop. The plans contain eight sheets of shop drawings and illustrated step-by-step directions. It's a totally professional piece of trade craft.

Now that I've reviewed his product plans, please read this carefully: Sardaukar Press does not sell finished items of any kind. Sardaukar sells plans and other published items, period. He also informs each purchaser of plans that an approved ATF Form 1 is required before any of these firearms is built and explains how to get such approval. He sells copies of the ATF Form 1.

"My original designs are examples of low-technology, highly concealable personal defense weaponry," Sardaukar said. "My plans are sold only for informational purposes. All plans and shop drawings are professionally drafted, carefully checked, and contain all information needed for completion, after official ATF approval is granted on a Form 1.

"As a registered professional engineer, I created my weapons as simple shapes using common materials to build with simple tools requiring little mechanical skill or experience. Using the same plans we sell, we had a legally built and tested prototype created for each weapon," he explained.

Sardaukar said he got his ideas for these improvised weapons as an undergraduate engineering student. He was watching the movie *Our Man Flint*, in which the hero, played by James Coburn, used a cigarette lighter that doubled as a gun.

"I knew I could make a real gun/lighter and quickly knocked out a sketch. That was it, until about five years later, when I came across that old sketch and decided to produce it in my very, very small and modest amateur workshop. I soon had a workable model," he explained.

After making careful inquiries of ATF, Sardaukar learned how much time, effort, and money were required to produce legal versions of his lighter/gun. So he destroyed his prototype and decided to sell only plans.

"We have a First Amendment, and you don't have to buy a government license to publish . . . yet . . . so I now sell only plans and have been doing so for about 10 years," he noted.

He added that the basic philosophy of Sardaukar Press is that his designs have to work, be relatively simple to make from common materials, and be defensive in nature. He is appalled at items of indiscriminate mass destruction such as bombs, grenades, mines, etc.

"I am always tinkering around in my mind for new ideas and new ways to hide working weapons in unusual forms. Right now I am working on an umbrella gun design," he added.

Perhaps Sardaukar is a student of the Kennedy assassination. If so, he might wish to read the next chapter.

10 CHAPTER TEN

Under the Umbrella

AT ITS BEST, AN UMBRELLA IS PORTABLE proof of our attempts to have civilization control environment. Sometimes, though, an umbrella may have more up its shaft than spokes and fabric, including an integral firearm.

Putting firearms in unusual and unexpected locations is part of the strategy of the professional shooter, a move designed to keep the target or the authorities off balance and to give the edge to the shooter. As I noted, guns go everywhere. Certainly, the use of umbrella guns, while novel, is hardly new.

In his excellent article "Hidden Thunder," noted arms historian David Fink writes about a fine black umbrella that is also a .36-caliber underhammer percussion rifle, the entire shaft of the umbrella being the weapon's barrel. It is marked with the year 1860. Fink describes another nineteenth-century umbrella gun where a small pepperbox revolver is part of the handle and is withdrawn from the shaft for firing.

In 1992, I saw two guns that were current modifications of regular umbrellas. Both were in police custody. One was a .410-gauge pistol whose 10-inch barrel was screwed into the remainder of the umbrella shaft. When

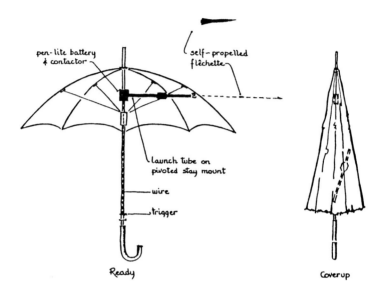

Diagram of the firing device modified into an umbrella and shooting a deadly dart, which "set up" JFK for his murder. (Courtesy of R.B. Cutler.)

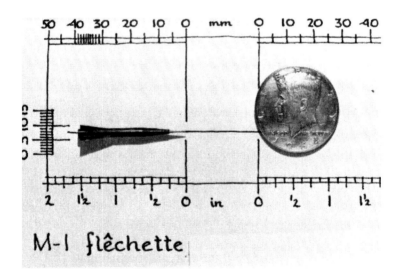

Graphic shows size of Mk1 flechette compared with Kennedy half dollar. (Courtesy of R.B. Cutler.)

Assassinology expert R.B. Cutler with a Mk1 flechette. (Courtesy of R.B. Cutler.)

the handle, receiver, and barrel were unscrewed, the weapon could be fired. The handle/barrel was held together by a screw, which had to be removed manually to eject the spent case and to reload. The other weapon was a .32-caliber rifle with the barrel comprising the entire length of the umbrella shaft. The police had confiscated both weapons in a raid on a drug distribution center.

But it was the major murder story of the twentieth century that brought the umbrella gun into real headline news. One of the highly reputable investigators looking for the truth of John Kennedy's murder has developed a theory that involves just such an improvised weapon. R.B. Cutler has published a book that presents the thesis that the so-called Dealey Plaza "umbrella man" was one of the JFK assassination team's members. The man's weapon was built into the open umbrella he was holding aloft as the Kennedy motorcade approached. That day was sunny, bright, and clear, so it does seem odd that this man would have his umbrella up and open, the only one in sight.

Cutler's theory holds that the umbrella contained a modified dart-firing gun which used very high-pressure propellant to fire poisoned projectiles. When the motorcade approached his position near the curb, "the umbrella man" opened his umbrella and held it aloft. After all the shooting, including his own silent shot, the man brought his umbrella down and melted into the confusion.

The actions of this man with the umbrella may be clearly seen in several still photos taken that day, but most clearly in the famous Zapruder film, which documents the actual assassination. What Cutler proposes is both technically and realistically possible. As for the actions of the "the umbrella man," these are already documented by film and by eyewitnesses.

Rather than fixate on the president's murder by triangulated ambush gunfire and the subsequent, continuing cover-up, let's examine the generic umbrella weapon developed for just such operations. Umbrella guns fall into the genre of improvised/modified weapons, and many have been produced professionally.

An earlier book showed a working model of a firearm built into an umbrella. It's been documented that both the OSS and CIA conducted experiments with and actively used lethal air-powered weapons built into various disguises, such as umbrellas. Until former CIA director William Colby revealed that the Agency actually had used exotic poisons, paralyzing agents, and weapons systems like this for assassination missions, people thought the idea was a flight of fiction.

In 1975, Colby and other CIA officials admitted to members of the U.S. Senate Select Committee on Intelligence that their R&D people had developed various disguised weapons—including an umbrella model—to fire a self-propelled rocket dart, which Colby called a flechette. The CIA people said the darts could carry either a paralyzing agent or fatal poison.

There is absolutely no doubt that the flechette-firing

umbrella was designed for the CIA. Senior CIA officials have testified to that, and it is now public record that the Agency's Health Alteration Committee, under the direction of Dr. Sidney Gottlieb and Col. Boris Pash, experimented with weapons systems that used drugs, esoteric poisons, and viruses, i.e., germ warfare. According to official testimony before Sen. Frank Church's Senate Select Committee on Intelligence in 1975, one of these weapons was described as "a noiseless, disguisable weapon firing darts which could be contaminated with LSD, germs, or venom."

In other testimony, Charles Senseney, a U.S. Army weapons expert, was asked by Sen. Howard Baker if the dart-firing assassination device could actually be used in the field in its disguised versions. Senseney replied evenly, "Oh yes, Senator, the M-1 projectile [the dart] could be fired from a cane; also from an umbrella. We've used both."

Senseney testified that he had developed the darts at Fort Detrick, Maryland, for CIA and for the army's Special Forces. The firing systems for the inch-and-a-half projectiles are nearly silent, smokeless, and recoilless. They are fired electrically using a solid-state fuel and have a velocity of 750 to 800 fps and an accurate and effective range of about 80 meters. The umbrella-disguised weapon used a battery-powered electrical circuit to ignite the dart's propellant and fire it through a straw-sized metal tube, which was mounted on the framework of the umbrella, attached to one of the ribs.

Cutler wrote: "The flechette looks like the tip of a large chicken feather. The needle nose is held in a block of plastic which has tiny tail fins for flight stability. These darts were first described to me by Col. L. Fletcher Prouty, a retired CIA liaison officer who knew of their development."

Later, Prouty told me, "I personally saw the flechettes at Fort Detrick on 29 July 1960, according to my records. I saw this tiny nylon or plastic rocket, a perfect miniature of about 5 millimeters diameter, with tail fins and solid propellant built into its body. They were obviously top-secret stuff.

"There were less than 10 CIA and Special Forces people who would have had personal access to that flechette and a launcher in 1960, and I doubt that number had grown to more than 30 by 1963. That is how highly classified and controlled these assassination devices were.

"That they were actually used to assassinate human targets would be debatable, because it would be tough to find hard evidence," Prouty added. "But there is lots of hard evidence about the existence of the Agency's secret assassination unit and the disguised weapons they used, like those deadly dart-firing umbrella guns."

According to a public statement made by former CIA official E. Howard Hunt in December 1975, the Agency had an in-house unit for the assassination "of suspected double agents and others . . . It had been operational since the mid-fifties." Hunt added that he'd been told that Col. Boris Pash was in charge of this unit. Senator Church's investigating committee found that this unit was known officially inside CIA as Program Branch 7 (PB/7). No records or documents about this unit existed by 1975, during the Church committee investigation of CIA-directed assassinations, other than one policy memo which stated in part, "PB/7 will be responsible for assassinations . . . and other such functions which . . . may be given it . . . by higher authority."

In his testimony to the Church committee, Senseney said that he and his associates produced the more exotic weapons used in some of this activity "out at the Branch unit." Did he mean Colonel Pash's PB/7 unit? Senseney refused to say. He claimed he didn't know what happened to his devices, who used them, or what they were used for.

Cutler has compelling evidence, though, that one of those devices was actually used in the murder of President Kennedy. Indeed, Cutler's research is such that he has written extensively on the assassination.

Considering the inherent danger of both the topic of his writing and the genre research, he might have considered the use of a pengun to compile his field notes. In the case of the pengun, truly, it is mightier than the sword.

11 CHAPTER ELEVEN

The Pen Is Mightier

THE MOST COMMON MODIFICATION of available hardware into firearm is that of the pengun. While a pengun is inaccurate, lacks muzzle velocity, cannot be reloaded quickly, and is often functionally unreliable, it can kill! And that effect is forever.

Because of cinematic and literary spy fiction, the pengun enjoys an armchair popularity. A pen or pencil is among the most common of common-folk, commonplace items. A pen's size, weight, and shape make it ideal for disguising a firearm of similar dimensions that would always be close at hand and ready for quick action.

Due to their small size, most penguns are .22 or .25 designs. Some of the commercial designs also come in .380 and 9mm. A few street modifications have shown up in .32 and .38, yet I saw a felt tip marker-sized model in Dade County, Florida, that was built around a .45 ACP round. It was not pleasant to fire. The firearms officer told me that one of his colleagues in nearby Miami had confiscated a similar-sized design that fired a .410 shotshell.

Sad experience with penguns has plagued their history. The wartime production item, known as the Stinger T2, probably caused more casualties to friendlies than to

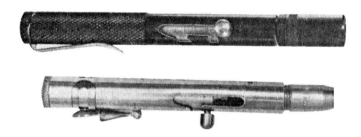

Examples of pen and tear gas guns, usually altered to fire fixed ammunition, depending on bore size. Common modifications convert the gas guns into .38- or .32-caliber firearms. Penguns are often .22-caliber. (Courtesy of ATF.)

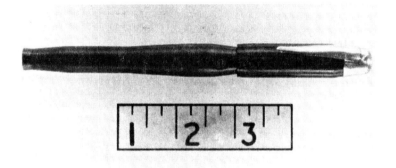

This assassin's weapon is a World War II vintage SOE pengun known as the SLAP model. It was used as a firing device to push against the victim in a stabbing or slapping fashion. (Courtesy of SOE archives.)

enemy personnel, according to the late Mitchell L. WerBell III, an OSS veteran and later exotic firearms impresario. WerBell once related an incident to me that occurred during a demonstration on the safe handling and use of the Stinger, where the OSS instructor took great pains to impress upon the agent trainees the proper handling of the weapon and promptly shot himself in the stomach with it.

The T2 Stingers were parachuted to the OSS- and SOE-led Kachin tribesmen in Burma during World War II. The tribesmen regarded them as magic totems because their method of handling them made it a toss-up as to whether the bullet would exit downrange or into the

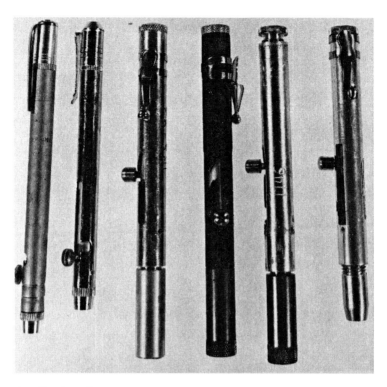

A collection of penguns seized by ATF after they had been converted to fire regular ammunition ranging from .22- to .32-caliber. (Courtesy of ATF.)

firer's own body, WerBell said. "They just couldn't get the idea into their head as to which direction the muzzle should point."

WerBell explained that the T2 was intended as a personal defense weapon, in that, like the FP45 Liberator, it was intended to be used to kill a Japanese soldier so the Kachin warrior/guerrilla could recover and use the Japanese military-issue weapons.

The colorful WerBell then related his only anecdote of seeing a Stinger in action: "We had a Japanese prisoner who was accidentally wounded by a careless damned Brit who was fooling around with a T2. He caused four holes in that poor prisoner with just one shot. He shot the man in the side of his ass cheek and got

Former ATF Director Rex Davis displays several penguns converted to fire .22-caliber ammunition. His bureau classified these various devices as firearms on 1 June 1975. (Courtesy of ATF.)

two entry and two exit wounds."

Think about it.

The checkered safety history of penguns fared no better after that war. Ironically, two of WerBell's employees and one of his sons also suffered self-inflicted wounds as a result of mishandling the modern SSSW Stinger produced by WerBell's Military Armaments Corporation in the late 1960s and early 1970s.

Most of the accidents are due to pure negligence, like placing fingers in front of the muzzle. The nonserious concern that some people who shoot large-caliber handguns hold for the tiny pengun results in their treating it unlike a "real" firearm. This careless stupidity leads to the "I didn't know it was loaded!" yelp of a hole-in-the-hand or lost fingers mishap.

Sometimes, penguns do more than just bite the hand that feeds them. In one case reported in the forensic literature in 1978, a conventional 6-inch pengun firing a .45 cartridge exploded, driving both the barrel and the bullet through the firer's eye and into his brain, causing a painful, lingering fatality. The weapon was a straight production pengun, not a modification of a tear gas device, i.e., it had both a serial number and caliber markings, and thus was obviously not homemade.

Commenting on the use of large-caliber handgun ammunition for small, lightweight penguns, the medical examiner in this case, Dr. D.W. Oxley, noted: "In a small, smooth tubular weapon, devoid of conventional grips and weighing only a few ounces, Newton's Third Law can, as it did in this case, easily convert the weapon itself into a lethal missile."

In Hong Kong, then-Maj. Richard Keogh reported on another sort of pengun problem. Late in 1975, he passed along a story of a Hong Kong restaurant employee having his life saved when a shot fired by a robber's small-caliber gun hit a

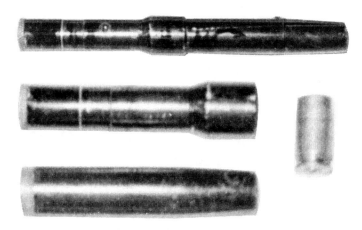

A modified .45 ACP pengun, shown assembled and disassembled, which malfunctioned, killing its owner. (Courtesy of David W. Oxley, M.D.)

steel fountain pen in the victim's pocket and ricocheted away.

Incidentally, according to a news report in a local newspaper, the robber was carrying "a modified toy pistol" that had been converted to fire live ammunition. Gun laws are very strict in Hong Kong, of course. There was no mention of whether the victim's pen was anything other than a writing instrument. As long as it saved his life, he probably did not care.

Penguns were never intended to be lethal weapons. They may have killed, but they were conceived, as the name "Stinger" implies, to sting, incapacitate, or distract the victim while something of a much more permanent nature is accomplished. Early manuals direct that the firer push his pengun at the target's face so the small-caliber round will cause a painful wound, and the gunpowder's flash will burn and blind his eyes, and the noise shock will disconcert him. The pengun is suited as a hidden, surprise, deceptive, or defensive weapon.

Penguns are usually striker-release weapons. For this reason most of the length of the pengun is taken up by the spring and bolt (striker) housing, leaving only a token barrel at best. The resulting inaccuracy demands that the weapon be fired at minimal ranges

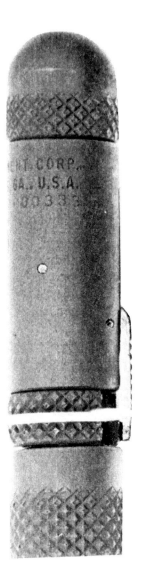

The lethal tube of self-protection, MAC's .22-caliber SSSW Stinger, used by police, Special Forces, and espionage agents. (Courtesy of Military Armament Corporation.)

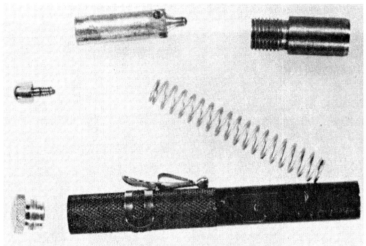

Several views of the Colt .25-caliber pengun involved in an accidental fatality to the firer. (Courtesy of Maj. Ralph Group.)

or, better yet, pressed directly against the face, throat, or base of the target's skull. Those are firearm penguns. Their smaller, slimmer, and more deadly cousins fire darts or needles, usually filled with toxins of varying use and lethality. These are almost exclusively designed, built, and used by professionals, e.g., intelligence agencies, terrorist assassins, and high-tech, freelance mercenary killers.

138 • Zips, Pipes, and Pens

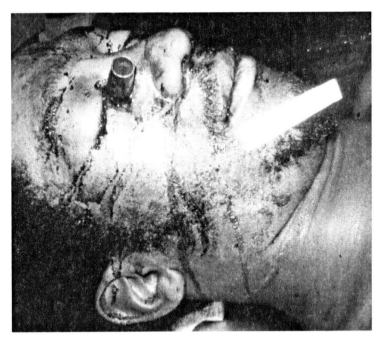

One of the dangers of modified weapons: the firer sometimes takes the projectile when the firing device, in this case a pengun, explodes.

These penguns are far more streamlined than the firearm models and use compressed gas or electrical ignition to launch their deadly missiles. These are intended to kill, which is why the CIA was "experimenting with" and then was discovered hoarding a quantity of cobra venom and shellfish toxin in the mid-1960s.

Improvised penguns are usually conversions of the tear gas and flare guns, e.g., the first British SOE penguns in World War II were tear gas guns to be used against potential captors or pursuing guard dogs. The simplest of these World War II devices was a two-part item consisting of a penlike body, which served as the barrel, and a cap, which was bolt and receiver with a fixed firing pin in its base. An agent simply slapped the base sharply with the flat of his hand, and the weapon fired.

World War II was also the era of the large fountain pen, which was obviously a metaphorical written invitation to the SOE and OSS designers to hide something more deadly than a poison-pen message. These fairly large OSS and SOE penguns, both in .22 and 9mm, were triggered by the pocket clip, a projecting button that turned the cap or released the striker from its L slot. Some designs were very clever, and many of them did, in fact, also write as a real pen should.

During one interview, an elderly World War II OSS vet said: "Those penguns which also wrote not only fooled the enemy, but we always felt that one might find a postwar civilian market among suicides who want to leave a farewell note written with the same device they will use to shoot themselves."

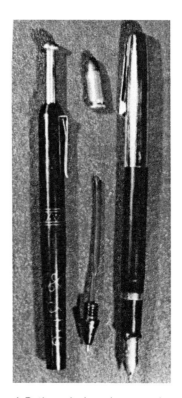

A Pathan-designed pengun in .25 ACP shown (left) next to an ordinary fountain pen (right). (Courtesy of Helms Tool & Die.)

Not surprisingly, modern technology gave post-World War II penguns a slimmer, more space-age look and capability. For example, as early as 1959, Soviet penguns fired with electrical resistance circuits and piezoelectric elements that ignited a powder charge to send the projectile on its way. These battery-fired penguns allowed more barrel length and a simple circuit-closing switch to actuate the firing.

As this book is written, though, the Cold War has been declared over, and the Soviet Union is no more. Now we all await the reaction of the International Professional War Machine to this concept that threatens its profits and prophets. However, war still continues in the streets of most of the world's cities, and penguns remain popular

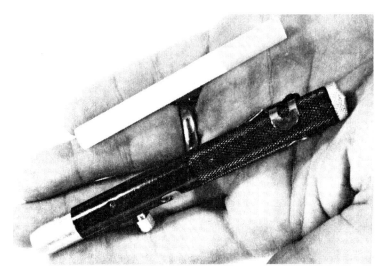

ATF official holds out a converted pengun and a cigarette for size comparison. (Courtesy of ATF.)

with these off-paper purchasers, a.k.a. street-gang hitters. According to New York Police Department (NYPD) annual crime statistical reports, in 1980 one of every 25 firearms confiscated was a pengun of some sort, ranging from the crude rubber band models to the well-machined high-tech finishes. By 1990, one of every 17 firearms was a pengun. A footnote added that by 1990, most were quality designs built by professional machinists.

In a 1989 order sent to all commands, the NYPD Police Commissioner's staff warned officers that at least one illegal gun factory was turning out a variety of professionally finished penguns in both .22-caliber and 9mm. The street price was reported to be only $60.

According to ATF regulations, penguns are registered as "any other weapons" and are transferred on a Form 4 with payment of a $5 transfer tax. Only Class III dealers can handle sales and transfers of these types of firearms, by the way. A regular FFL license does not qualify an individual to transact business in this category of firearm. ATF would like laws to ban the possession of penguns altogether, according to most agents with whom I've spoken.

Passing more laws against penguns will not stop their use any more than confiscation would. Their basic materials and components are so common that they will always be with us. There is no simple answer to the problem of firearms control and public safety, and obviously no hardline answers will solve the problem either. Meanwhile, commercial production goes on.

In the early 1980s, Covert Arms Manufacturing produced three configurations of the pengun for licensed commercial sale, including .22 Long Rifle and .22 Magnum models, plus a .410 shotgun version. The .22 caliber models were 3/8" x 4 3/4" and 1/2" x 5", while the .410 was a rather robust 3/4" x 9 1/4" long. List prices for the three weapons ranged from $73.50 through $76 for the .22 models to $89.75 for the shotgun version.

Their finish was anodized and, unlike some penguns, each had a positive safety device to prevent accidental discharge. A pen-clip device was included for shirt- or jacket-pocket carry. These firearms were advertised both as professional trade craft backup weapons and also as valuable and rare firearms for the serious collector.

In 1992, Long Mountain Outfitters of Harmony, Maine, was offering Hornet penguns for sale commercially. The Hornet was produced in .22, .25, .380, and 9mm by Armitage International, exclusively for distribution by Long Mountain. They billed it as a survival tool for field use. They also suggested that this versatile firearm can double in occupation as a kubotan. The Hornet is 6 inches in overall length and is 3/4 inch in diameter. According to LMO advertising, each .22, .380, and 9mm barrel is threaded for a sound suppressor.

Johnny Vitte is another producer of such arcane weaponry as the pengun. A Florida-based armorer, Vitte produces a meticulously engineered .22 pengun. Working from what he called "a tinkerer's hokey-pokey, trial-and-error design," he gradually created what became his production model, a combination of "real gun parts, real pen parts, and some homemade fabrications." His pengun will accept a shortened standard ballpoint pen cartridge and actually writes. The finish is also an interesting point.

Three views of the Vitte "Power Pen."

"I figured the people at Parker Pens, who make some very fancy writing instruments, didn't parkerize their pens, so why should I," said Vitte with a laugh. "I finish each unit with a thin coat of sprayed black polyurethane paint that gives each that real expensive pen look and feel."

The Vitte pengun works about the same as most of the others, consisting of a spring and striker. Vitte said firing what he calls his "Power Pen" is simply a matter of removing the pen cap and writing tip while holding the unit in your hand pointed downrange. Use the thumb of that same hand to slide the cocking knob to the rear, which, of course, brings back the bolt, suppressing the spring.

Firing is a simple matter of releasing the cocking knob so that the spring forces the bolt forward, bashing it against the primer, which fires the round. Vitte added that the recoil from the light .22 Long Rifle round is virtually nonexistent, and the sound level within that closed bolt system is about like that of a starter pistol.

Penguns remain popular, prevalent, and in production, both legally and otherwise. They are controversial subjects that are going to remain with us.

On the lighter side of the pengun issue, perhaps a careless enforcer or someone with a bizarre twist of humor left behind a deadly calling card in a Quebec courtroom. Checking the court facility after the first day's hearings in an organized crime investigation, security people found that one of the spectators had dropped a somewhat heavy but totally functional felt-tip marking pen on the floor and failed to pick it up. While checking it into lost and found, the security man fumbled with the pen's clasp, which triggered a .22 cartridge, driving the bullet into the ceiling. Yes, it was the old pengun-in-the-courtroom trick. Although security was tightened the following day, the point was not lost. It surely was not lost on the security man, who found this well-disguised improvised firearm and nearly got the last surprise of his life.

Bibliography

I'D LIKE TO ACKNOWLEDGE THE HELP of many friends and other sources in locating, compiling, and presenting the technology, histories, anecdotes, data, and photographs in this book. They include Frank C. Brown; David H. Fink; Harold Johnson, formerly of the U.S. Army's Foreign Science & Technology Center; Maj. Richard Keogh; Jack Krcma; John Minnery; Ed Owen, the ATF's resident maven on the topic; William Randall; Fred Rexer, Jr., warrior turned Hollywood creative genius; author-designer Richard Sardaukar; plus old friends Peter Senich, Pat Taylor, and Donald G. Thomas.

In addition, I owe many thanks to various law enforcement officers and organizations, as well as military and intelligence officers and organizations nationally and internationally for their help in making this book possible.

There is also this informal bibliography for those readers who seek further technical and historical information:

Ahern, Jerry. "The Gun—Ultimate Symbol of Self-Reliance." *Guns & Ammo* (June 1980).
Avery, Ralph. *Combat Loads for the Sniper Rifle*. Cornville, AZ: Desert Publications, 1981.

The AR-7 Advanced Weapons System. Boulder, CO: Paladin Press, 1986.
Benson, Ragnar. *Breath of the Dragon*. Boulder, CO: Paladin Press, 1980.
Brown, Ronald B. *Homemade Guns and Homemade Ammo*. Port Townsend, WA: Loompanics, Inc., 1986.
Burns, Sgt. G.T. *Improvised Weapons*. Independently published, 1971.
Cutler, R.B. *The Day of the Umbrella Man*. Manchester, MA: Cutler Designs, 1980.
Davidson, Bill R. *To Keep and Bear Arms*. Boulder, CO: Paladin Press, 1979.
DuPuy, Trevor. *European Resistance Movements*. New York: Franklin Watts, 1965.
Elias, Stephen. *Legal Research*. Berkeley, CA: Nob Press, 1983.
Fink, David H. "Hidden Thunder." *Man at Arms* (Jan./Feb. 1989).
Gottlieb, Alan M. *The Rights of Gun Owners*. Aurora, IL: Caroline House Publishers, Inc., 1981.
Grennell, Dean. *The ABCs of Reloading*, 4th ed. Northfield, IL: DBI Books, 1988.
Gudz, Rob. ".22 Caliber Pen Gun." *Machine Gun News* (Jan. 1992).
Hatcher, Maj. Gen. Julian. *Hatcher's Notebook*. Harrisburg, PA: Stackpole Books, 1947.
Hayduke, George. *The Hayduke Silencer Book*. Boulder, CO: Paladin Press, 1989.
Hermann, Richard L. "Asiatic Pistols." *Auto Magazine* (March 1970).
____."Asiatic Pistols, Part 2." *Auto Magazine* (October 1972).
Hogg, Ian V. *Guns and How They Work*. New York Everest House, 1979.
Holmes, Bill. *Home Workshop Guns for Self-Defense and Resistance*. Boulder, CO: Paladin Press, 1977.
Improvised Munitions Handbook. Philadelphia, PA: Frankford Arsenal, undated.

Karnoff, Kingsley. "Percussion Machine Gun." *Guns* (Dec. 1965).

Koch, Robert. *FP45 Liberator Pistol*. Long Beach Research, 1975.

Kukla, Robert J. *Gun Control*. Harrisburg, PA: Stackpole Books, 1973.

Levy, Bert. *Guerrilla Warfare*. Boulder, CO: Paladin Press, 1968.

Long, Duncan. *The AR-15/M16: A Practical Guide*. Boulder, CO: Paladin Press, 1990.

———. *AR 15/M16 Super Systems*. Boulder, CO: Paladin Press, 1988.

———. *Combat Ammunition*. Boulder, CO: Paladin Press, 1989.

———. *Making Your AR-15 into a Legal Pistol*. Boulder, CO: Paladin Press, 1986.

———. *Mini-14 Super Systems*. Boulder, CO: Paladin Press, 1985.

"Machinegun Conversion of the U.S. .30 Carbine." Service Sales Co., 1964.

Matunas, Ed. "All-Purpose 12-Gauge." *Handloader* Number 93 (Sept./Oct. 1981).

Meadows, Capt. E.S. "Weapons of the Vietcong." *Guns & Ammo* (April 1969).

The Mini-14 Exotic Weapons System. Boulder, CO: Paladin Press, 1985.

Minnery, John A. and Joe Ramos. *American Tools of Intrigue*. Cornville, AZ: Desert Publications, 1980.

Minnery, John. *How to Kill, Vol. 5*. Boulder, CO: Paladin Press, 1980.

———. *Fingertip Firepower*. Boulder, CO: Paladin Press, 1989.

———. "The O.N.I. Fist Gun." *The Gunrunner* (Sept. 1973).

———. "The OSS Liberator." *The Gunrunner* (Sept. 1972).

Most, Johann. *Military Science for Revolutionaries*. Cornville, AZ: Desert Publications, 1978.

———. *Improvised Weapons of the American Underground*. Boulder, CO: Paladin Press, Reprint 1975.

Powell, William. *The Anarchist Cookbook*. Secaucus, NJ: Lyle Stuart, Inc., 1979.
Rees, Clair F. "Underground Arsenals behind Prison Walls." *Guns & Ammo* (Jan. 1970).
Rideout, Granville N. *The Chicom Series*. Ashburnham, MA: Yankee Publishing Co., 1971.
Riste, Olav and Bent Nokeby. *The Resistance Movement: Norway*, 1940–1945. Oslo: Johan Grundt Tanum Forlag, 1970.
Rousell, Aage. T*he Museum of the Danish Resistance Movement*. Copenhagen: The National Museum, 1968.
Ruger 10/22 Exotic Weapons System. Boulder, CO: Paladin Press, 1985.
Saxon, Kurt. *The Poor Man's James Bond*. Harrison, AR: Atlan Formularies, 1972.
Smith, Leslie L. "Zip Guns." *Police* (Jan./Feb. 1963).
Smith, Mark. *Hidden Threat*. Boulder, CO: Paladin Press, 1990.
Storm, Bill. "Guns of Riots." *Guns* (Nov. 1972).
Tappan, Mel. *Survival Guns*. Rogue River, OR: Janus Press, 1980.
Thornbrugh, Wayne. *Select Fire 10/22*. Boulder, CO: Paladin Press, 1989.
Truby, J. David. *How Terrorists Kill*. Boulder, CO: Paladin Press, 1979.
____. "More Quiet Killers." *Special Weapons Annual* (Jan. 1983).
____. "New Police Weapon Report." *Police Intelligence Newsletter*, 1973.
____. *The Quiet Killers*. Boulder, CO: Paladin Press, 1972.
____. *Silencers, Snipers & Assassins*. Boulder, CO: Paladin Press, 1972.
____. "Stinger." *Overseas Weekly* (Jan. 15, 1973).
Typical Foreign Unconventional Warfare Weapons. Charlottesville, NC: U.S. Army Foreign Science & Technology Center, 1964.

U.S. Army. *Improvised Munitions Handbook* TM 31-210. Philadelphia, PA: Frankford Arsenal, 1969.

Wells, Robert. *The Anarchist Handbook*. Miami: T. Flores Publications, 1985.

Whisker, James B. *Our Vanishing Freedom*. Skokie, IL: Publishers Development Corp., 1972.

Winant, Lewis. *Firearms Curioso*. Philadelphia, PA: Ray Riling, 1955.

Sources

HERE ARE SOME BUSINESSES THAT are knowledgeable about improvised and modified firearms or who have designs, plans, parts, or various other cogent items for sale. At the time this book was published, each of these was a viable, operational business. You should also consult current issues of such periodical trade publications as *Gun List*, *Small Arms Review*, and *Shotgun News* for updated and/or additional information.

DELTA GROUP
215 S. Washington
El Dorado, AR 71730
http://www.deltaforce.com

GUN PARTS CORPORATION
226 Williams Lane
West Hurley, NY 12491
http://www.gunpartscorp.com

LONG MOUNTAIN OUTFITTERS LLC
631 N. Stephanie St., #560
Henderson, NV 89014
http://www.longmountain.com

LOOMPANICS UNLIMITED
P.O. Box 1197
Port Townsend, WA 98368
http://www.loompanics.com

PALADIN PRESS
Gunbarrel Tech Center
7077 Winchester Circle
Boulder, CO 80301
http://www.paladin-press.com

SARCO, INC.
323 Union St.
Stirling, NJ 07980
http://www.sarcoinc.com

SPRINGFIELD SPORTERS, INC.
2257 Springfield Road
Penn Run, PA 15765
http://ssporters.com

Also, I would be pleased to hear from readers who have photos, news clippings, accounts, reports, or other data, textual and graphic, related to improvised/modified firearms that would be appropriate for a future book on this topic. Please send your information to me:

J. David Truby
c/o Paladin Press
Gunbarrel Tech Center
7077 Winchester Circle
Boulder, CO 80301

I will answer each letter personally.

About the Author

J. DAVID TRUBY is a former editor for National News Service (NNS), an international news and feature syndicate. He covered political, military, intelligence, terrorism, paramilitary, and assassination. Truby also covered Latin America for NNS.

Dr. Truby is a member of Investigative Reporters and Editors, Inc., Reporters Committee for Freedom of the Press, and the Society of Professional Journalists.

He has written seventeen books and coauthored five others. He has free-lanced more than 1,200 major magazine, newspaper, and television stories in addition to his other work. His reporting, photojournalism, and writing have won seven national awards. He has worked in magazine, newspaper, radio, and television news. A U.S. Army veteran, Dr. Truby was a combat intelligence NCO. He has taught criminology, English, and journalism at the university level.

At present, Dr. Truby is involved as an armament consultant in both domestic and international markets.